AF598096

NEW TRENDS IN NUCLEAR SCIENCE

Edited by **Nasser Sayed Awwad**
and **Salem A. AlFaify**

New Trends in Nuclear Science
http://dx.doi.org/10.5772/intechopen.74762
Edited by Nasser Sayed Awwad and Salem A. AlFaify

Contributors

Andres Rodriguez-Hernandez, Armando Miguel Gomez-Torres, Edmundo del Valle Gallegos, Thomas W. Grimshaw, Olena L. Maslyanchuk, Stepan Melnychuk, Volodymyr Gnatyuk, Toru Aoki, Joji M. Otaki, Nasser Sayed Awwad, Salem A. AlFaify

First published in London, United Kingdom, 2018 by IntechOpen
IntechOpen is the global imprint of INTECHOPEN LIMITED, registered in England and Wales, registration number: 11086078, The Shard, 25th floor, 32 London Bridge Street
London, SE19SG – United Kingdom
Printed in Croatia

British Library Cataloguing-in-Publication Data
A catalogue record for this book is available from the British Library

Additional hard copies can be obtained from orders@intechopen.com

New Trends in Nuclear Science, Edited by Nasser Sayed Awwad and Salem A. AlFaify
p. cm.
Print ISBN 978-1-78984-656-0
Online ISBN 978-1-78984-657-7

For EU product safety concerns:
IN TECH d.o.o., Prolaz Marije Krucifikse Kozulić 3, 51000 Rijeka, Croatia,
info@intechopen.com or visit our website at intechopen.com.

Meet the editors

Dr. Nasser Sayed Awwad received his MSc in Inorganic and Radiochemistry in 1997 from Benha University, his PhD in Inorganic and Radiochemistry in 2000 from Ain Shams University and Post doctorate at Sandia National Labs, New Mexico, USA in 2004. Nasser Awwad was Associate Prof of Radiochemistry in 2006 and professor of Inorganic and Radiochemistry in 2011 at the Egyptian Atomic Energy Authority. He has been a Professor at King Khalid University, Abha, KSA, from 2011 to the current day. He has published two chapters in the following books: "Natural Gas - Extraction to End Use" and "Advances in Petrochemicals". He has been Editor for four books: "Uranium", "New trends in Nuclear Sciences", "Dyes in Industry" and "Lanthanides". In addition to this, he has published 54 papers in ISI Journals. He has supervised 4 PhD and 16 MSc students in the field of Radioactive and Wastewater treatment. He has participated in 25 International conferences (South Korea, USA, Lebanon, KSA and Egypt). He has reviewed 2 PhD and 13 MSc papers. He has participated in 6 big projects with KACST at KSA and Sandia National Labs at USA on the conditioning of radioactive sealed sources, wastewater treatment. He is the leader of group research about utilization of nanomaterials for treatment of inorganic and organic pollutants and is a member of a research group on the influence of d- and f-block element interactions with indomethacin, bupivacaine, lidocaine, nadolol, procainamide and procaine nonsteriodal drugs on its bioactivities. He is a member of the Arab Society of Forensic Sciences and Forensic Medicine and a member of the Egyptian Society for Nuclear Sciences and its applications. He is part of the Editorial Board of the Journal of Energy and Environmental Research and Technology. He is also a member of the organizing committee of the 9th World Congress & Expo on Nanotechnology & Material science and a rapporteur of the Permanent Committee for Nuclear and Radiological Protection at King Khalid University. He is interested in utilizing the different techniques related to treatment of radioactive nuclides using nanomaterials.

Salem A. AlFaify (S. AlFaify) is currently an associate professor of physics, leader of the "quantum functional materials for advanced applications" (QFMAA) research group and a leading researcher at the advanced functional materials and optoelectronics laboratory (AFMOL) at the department of physics faculty of sciences, King Khalid University (KKU). He was President of the Saudi Physical Society (SPS) from 2013-2016.

He obtained his Ph.D. in condensed matter physics/nano-materials in 2011 from the Western Michigan University, USA. He was awarded a thesis appointment scholarship from the department of energy DOE-USA to conduct his Ph.D research project in the center for nano-scale material (CNM) at Argonne National Laboratory (ANL), one of the dominant national laboratories of the DOE-USA operated and managed by the University of Chicago. He has authored and co-authored more than 150 articles in peer-reviewed and highly reputed ISI journals. He works with collaborators/researchers with mutual interest from many institutes and universities around the world. His research interests are primarily in the area of the condensed matter physics at the nano-scale, in particular, the correlation nature of the nano-quantum structures and their properties and applications. In addition to the growth of varying forms of nanostructured materials and their basic characterization by XRD, SEM, TEM etc., he is interested in utilizing the powerful techniques related to accelerators physics such as ion beam analysis (IBA) and synchrotron radiation to fundamentally investigate the essence of the nanomaterials/fine matters and understand/engineer their novel properties for modern and futuristic applications.

Contents

Preface

The book deals with topics related to the new trends of research in nuclear science and its subjects. It contains five chapters. The first one is an introductory chapter to explain the nature and purpose of the book and the logic and significance of its contents. The second chapter is a concise introduction to the core subject of nuclear science, which is the nuclear reactions. Chapter three addresses some recent advances related to the famous nuclear detector material namely CdTe. Another new trend of nuclear science research mentioned in this book is contained in chapter four. Joji M. Otaki has investigated the multifaceted biological effects from the tragic Fukushima nuclear accident caused by mother nature to help researchers understand the correlation of environmental effects on low-dose exposure. In the last chapter, Thomas W. Grimshaw composed an interesting study on the so-called cold nuclear fusion. The book presents recent knowledge on frontier topics of nuclear science research for chemists, physicists and engineers alike working in the area of nuclear reactors and reactions, applied nuclear materials, and environmental effects. The book provides valuable insights into some modern topics related to nuclear technology and its applications in our daily life. Finally, the book's editors would like to express their gratitude and appreciations to the book contributors and IntechOpen Author Service Manager. We hope this book will be a useful addition to the rich body of knowledge available on modern nuclear science and we look forward to a brighter future from its applications and technologies.

Nasser Sayed Awwad
Chemistry Department, Faculty of Sciences
King Khalid University
Abha, Saudi Arabia

Salem A. AlFaify
Physics Department, Faculty of Sciences
King Khalid University
Abha, Saudi Arabia

Chapter 1

Introductory Chapter: Introduction to New Trends in Nuclear Science

Salem A. AlFaify and Nasser S. Awwad

Additional information is available at the end of the chapter

http://dx.doi.org/10.5772/intechopen.82231

1. Introduction

Nuclear energy is still one of the most thriving technological sectors in our modern life. It is considered by far one of the leading sources of alternative energy. That modern energy many driving forces worldwide are looking for, since, it does not share limitations and setbacks of the current energy sources based on fossil materials which are costly, depletable, and most importantly not environmental friendly. In a recent major interdisciplinary study from the "MIT Future of Series" initiative, a comprehensive report on the future of nuclear energy in a carbon-constrained world was published, suggesting strongly that nuclear energy is a potential solution to the challenge facing our modern world to establish sustainable mechanisms for producing energy that is based on low carbon technologies.

Many researchers, scientists, and engineers alike on the fields of nuclear science have come with similar conclusions of the MIT study on nuclear energy and insist that more efforts are needed to lower the cost of nuclear technology and optimize its applications in our daily life at the same time. Therefore, many research groups on nuclear science from around the globe are actively working in a variety of fronts to expand the body of knowledge available on nuclear materials and explore new ways for their utilization in modern and futuristic applications. Hence, this book is going to shed light on some of the advances taking place within nuclear science research in recent time. It is a small effort to show interesting results of some modern nuclear science research carried out by bright scientist and research in different parts of the world.

The book is divided into five chapters. The first one is a short introduction to explain the nature and purpose of the book and the logic and significance of its contents. The second one is a concise introduction to the core subject of nuclear science, which is nuclear reaction. The chapter

touches lightly on the fundamental and basic physics underlining major nuclear reactions, i.e., nuclear fissions. In addition, the chapter shows different generations of nuclear reactors as important tools to harness energy resulting from nuclear reactions. Some of these reactors are relatively modern and hold great potentials for future nuclear technology. The second chapter comes with many figures and charts along with two appendices to point out many aspects of nuclear fission reactions and their utilization in modern nuclear reactors.

In the third chapter, attention is to direct to another important field of nuclear science that is nuclear detectors. To be more precise, the third chapter addresses some recent advances related to the famous nuclear detector material namely CdTe. Maslyanchuk et al. [1], suggest that the modern detector based on CdTe materials can be developed as multielement detection platform that allows for the direct conversion of information generated by passing X/γ-radiations through an examined object into an array of digital electrical signals without using an intermediate visible image on a fluorescence screen. Such an approach will facilitate real-time visualization and sufficiently enhance image resolution. In the chapter, discussion on different aspects of the semiconductor nuclear detector based on CdTe material and Schottky effect shows progressive research into one of the modern nuclear detection devices that will enable developers to utilize nuclear technology in safe and practical ways [2, 3].

In the next chapter, a new study on the effect of unintended and accidental nuclear impact on the environment is discussed. Recently, Joji M. Otaki, from the department of chemistry, biology, and marine science at the faculty of science of the University of Ryukyus, Japan, has investigated the multifaceted biological effects from the tragic Fukushima nuclear accident caused by Mother Nature. Such study may help researchers to understand the correlation nature of low-dose exposure and field effects on the environment and may eventually lead to resolving the field-laboratory paradox on the environmental damage caused by the low-dose radiation. The study concluded that the "low-dose" exposure from the Fukushima nuclear accident imposed potentially non-negligible toxic effects on organisms including butterflies and humans through environment field effects. At the high-dose exposure, same field effects can exist, but would likely be masked by the acute radiation damage on the subjected environment [4].

In the last chapter, Thomas W. Grimshaw, from the University of Texas at Austin, USA, has composed an interesting study on the so-called cold nuclear fusion or as widely known the low-energy nuclear reaction (LENR). He, among others, argued that nuclear cold fusion if realized and understood could be a significant source of cheap and clean energy. When LENR was introduced to the scientific communities in early 1989, it was greatly dismissed by the mainstream scientists. However, as the chapter's author mentioned, despite such rejection of LENR concept, the research activities on the subject are growing by the time, and the matter of LENR is still alive and vibrant. According to many researchers around the world, there is accumulating evidence on the reality of LENR phenomena. Such advances on the LENR subject should encourage interested communities and potential stakeholders to come up with policymaking and regulations to facilitate the growing research endeavors on LENR for its realization and harnessing its great benefits. Moreover, it is important to work on mitigating its secondary effects since LENR is expected to evolve into disruptive technology. Author has concluded that due to the recent updates, the support for LENR developments and preparations to mitigate its anticipated adverse secondary impacts is largely needed [5, 6].

http://dx.doi.org/10.5772/intechopen.82231

Although the presented book does not provide a comprehensive treatment by any means to its topics, it is still a very constructive venue to direct readers' attention to some of the advanced trends of nuclear science research. This book will definitely encourage readers, researchers, and scientists to look further into the frontier topics of modern nuclear science and make the needed efforts to develop its cause and uses.

Author details

Salem A. AlFaify[1]* and Nasser S. Awwad[2]

*Address all correspondence to: saalfaify@kku.edu.sa

1 Physics Department, Faculty of Sciences, King Khalid University, Abha, Saudi Arabia

2 Chemistry Department, Faculty of Sciences, King Khalid University, Abha, Saudi Arabia

References

[1] Kosyachenko LA, Maslyanchuk OL, Motushchuk VV, Sklyarchuk VM. Charge transport generation-recombination mechanism in Au/n-CdZnTe diodes. Solar Energy Materials and Solar Cells. 2004;**82**(1-2):65-73

[2] Szeles C. CdZnTe and CdTe materials for X-ray and gamma ray radiation detector applications. Physica Status Solidi B. 2004;**241**(3):783-790. DOI: 10.1002/pssb.200304296

[3] Sordo SD, Abbene L, Caroli E, Mancini AM, Zappettini A, Ubertini P. Progress in the development of CdTe and CdZnTe semiconductor radiation detectors for astrophysical and medical applications. Sensors. 2009;**9**:3491-3526. DOI: 10.3390/s90503491

[4] Levi G et al. Indication of Anomalous Heat Energy Production in a Reactor Device Containing Hydrogen Loaded Nickel Powder. Cornell University Library. arXiv. 2013. https://arxiv.org/ftp/arxiv/papers/1305/1305.3913.pdf

[5] Steen TY. Ecological impacts of ionizing radiation: Follow-up studies of nonhuman species at Fukushima. Journal of Heredity. 2018;**109**:176-177

[6] Møller AP, Mousseau TA. Strong effects of ionizing radiation from Chernobyl on mutation rates. Scientific Reports. 2015;**5**:8363

Chapter 2

Nuclear Reactor Simulation

Andrés Rodríguez Hernández,
Armando Miguel Gómez-Torres and
Edmundo del Valle-Gallegos

Additional information is available at the end of the chapter

http://dx.doi.org/10.5772/intechopen.79723

Abstract

A summary is described about nuclear power reactors analyses and simulations in the last decades with emphasis in recent developments for full 3D reactor core simulations using highly advanced computing techniques. The development of the computer code AZKIND is presented as a practical exercise. AZKIND is based on multi-group time dependent neutron diffusion theory. A space discretization is applied using the nodal finite element method RTN-0; for time discretization the θ-method is used. A high-performance computing (HPC) methodology was implemented to solve the linear algebraic system. The numerical solution of large matrix-vector systems for full 3D reactor cores is achieved with acceleration tools from the open-source PARALUTION library. This acceleration consists of threading thousands of arithmetic operations into GPUs. The acceleration is demonstrated for different nuclear fuel arrays giving extremely large matrices. To consider the thermal-hydraulic (TH) feedback, several strategies are nowadays implemented and under development. In AZKIND, a simplified coupling between the neutron kinetics (NK) model and TH model is implemented for reactor core simulations, for which the TH variables are used to update nuclear data (cross sections). Test cases have been documented in the literature and demonstrate the HPC capabilities in the field of nuclear reactors analysis.

Keywords: HPC, high-performance computing, NFEM, nodal finite element method, parallel computing, GPU, graphics processing unit, NK-TH, neutronic-TH coupling

1. Introduction

The mathematical models representing the nuclear reactor physics are based mainly on two theoretical areas: neutron transport theory and neutron diffusion theory, where it is necessary to remark that neutron diffusion theory is really a simplification of the neutron transport theory.

Numerical methods are used to solve the partial differential equations representing the nuclear reactor physics, and these methods are derived from discretization techniques. For numerical solutions in any scientific area, computational tools have been developed including software and hardware. In the past, the former computer processing was the sequential execution of computer commands, meaning to say that program tasks are carried out one after one. Modern computational tools have been developed for parallel processing, executing several tasks concurrently.

The computing branch dealing with the system architecture and appropriate software related to the simultaneous execution of computer instructions and applications is known as parallel computing science. Former developments in parallel computing were made in the late 1950s, following the construction of supercomputers throughout the 1960s and 1970s. Nowadays, clusters are the workhorse of scientific computing and are the dominant architecture in data centers.

Since the late 1950s, the performance of safety analyses was essential in the nuclear industry, in research reactors, but mainly safety analyses of nuclear power plants for commercial purposes. Scientific computing calculations were vital to these safety analyses, but with important limitations in computer/computing capabilities. At the beginning, the objective was to give a solution to partial differential equation models based on neutron diffusion or neutron transport with technology and methods available in those years. Numerical techniques were used first with finite differences and finite element approaches, and gradually up to now, with nodal finite element methods (NFEMs). Despite the numerical method employed, the computer code user faces the problem of solving extremely large algebraic systems challenging hardware/software capabilities. Generation of results for any reactor simulation in considerable short times is a desirable achievement for computer code users [1].

Recent developments of high-performance computer equipment and software have made the use of supercomputing in many scientific areas possible. The appropriate selection of parallel computing software, like newly developed linear algebra libraries, to be used in a specific project may result in a suitable platform to simulate nuclear reactor states with relatively prompt results.

Throughout the world, several research projects in the last decade have been developed with the main objective of making full tridimensional (3D) coupling simulations of nuclear reactor cores, leaving aside the obsolescence of the point kinetics theory. Most of the modern nuclear reactor simulators are based on neutron transport theory, or on neutron diffusion theory, to obtain detailed 3D results. As light water is used for cooling/moderating light water reactors (LWRs), a comprehensive analysis of the reactor core physics must include thermal-hydraulic phenomena, so that modern simulations perform reactor calculations with thermal-hydraulic feedback coupled with neutron kinetics calculations.

http://dx.doi.org/10.5772/intechopen.79723

All the discussions included in this chapter are centered in a simulator for light water reactors. The computer code AZtlan KInetics in Neutron Diffusion (AZKIND) is part of the neutronic codes selected for their implementation in the AZTLAN Platform[1] project in which neutron transport and neutron diffusion codes are being developed in Mexico. A (TH) model has been implemented recently and coupled with the neutronic (NK) model, and both models are based on HPC implementations.

2. Reactor core calculation overview

Although there has been growing interest in the transport-based core neutronics analysis methods for a more accurate calculation with high-performance computers, it is yet impractical to apply them in the real core design activities because their performance is not so practical on ordinary desktop or server computing machines. For this reason, most of the neutronics codes for reactor core calculations are still subject to the two-step calculation procedure, which consists of (1) homogenized group neutron parameters generation and (2) neutron diffusion core calculation.

In the core calculation steps that are the main concern of this work, nodal codes based on the diffusion theory have been used to determine the neutron multiplication factor and the corresponding core neutron flux (or power) distribution. Practically, almost all nuclear reactor simulation codes employ the two-group approach involving only fast and thermal neutron energy groups for the applications to light water reactors (LWR). However, numerical calculations with the two-group structure are not appropriate in the analysis of cores loaded with mixed oxide fuels or analysis of fast breeder reactors, since the neutron spectrum is influenced more by the core environment, requiring much more energy groups than only two groups.

As settled in Ref. [2], even using a high-performance computer, a direct core calculation with several tens of thousands of fuel pins is difficult to perform in its heterogeneous geometry model form, using fine groups of a prepared reactor cross-section library. The Monte Carlo method can handle such a core calculation (see also the Serpent code), but it is not easy to obtain enough accuracy for a local calculation or small reactivity because of accompanying statistical errors, besides the large calculation times. Instead of using neutron transport computer codes, the nuclear design calculation is performed in two steps: (1) lattice calculation in a two-dimensional infinite arrangement of fuel rods or assemblies for the generation of homogenized lattices jointly with their corresponding homogenized cross-sections and (2) core calculation in a three-dimensional whole core, with a neutron diffusion code using the information of the previous step.

As shown in **Figure 1** [2], the lattice calculation prepares few-group homogenized cross sections which maintain the energy dependence (neutron spectrum) of nuclear reactions, and these reduce the core calculation cost in terms of time and memory. The final core design

---

[1]This work was performed under the auspices of the financial support from the National Strategic Project No. 212602 (AZTLAN Platform) as part of the Sectorial Fund for Energetic Sustainability CONACYT—SENER, Mexico.

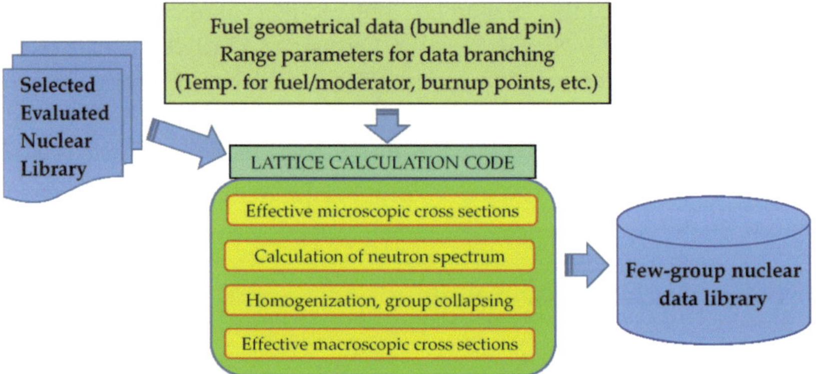

Figure 1. Typical lattice calculation process flow for light water reactors [2].

parameters are not concerned with continuous energy dependence, but spatial dependence, such as power distribution, is important to avoid high local neutron fluxes or high absorbing materials causing significant neutron flux gradients, mainly when safety analyses are performed upon the final proposed core designs.

In the core calculations with space-dependent data (cross sections and neutron flux), the effective cross sections are processed, with a little degradation in the accuracy as possible, by using the results from the multi-group lattice calculation. Lattice code calculation and codes are not discussed here.

There are two processes followed for lattice calculation. One is the homogenization to lessen the space-dependent information and the other is group-collapsing to reduce the energy-dependent information as shown in **Figure 2**. The fundamental idea of both methods is to preserve neutron reaction rate. The next step is to consider the conservation of reaction rate in the energy group G in the same manner as that in the homogenization.

The number of few groups depends on reactor type and computation code. Two or three groups are adopted for the NK- and TH-coupled core calculation of LWRs and much more groups (18, 33, etc.) are used for the core calculation of LMFRs (Liquid Metal Fast Reactors). Currently, revised methods exist for the improvement of cross-sections generation using computer codes dedicated to lattice calculation for few-groups approach, like in Ref. [3], where three topics are involved: (1) improved treatment of neutron-multiplying scattering reactions; (2) group constant generation in reflectors and other non-fissile regions, leading to the use of discontinuity factors in neutron diffusion codes; and (3) homogenization in leakage-corrected criticality spectrum, in which several leakage corrections are used to attain criticality, accounting for the non-physical infinite-lattice approximation. Another improvement was done in Monte Carlo codes [4], implementing reliable multi-group cross-sections calculations for collapsed flux spectrum. Ref. [4] focuses on calculating scattering cross sections, including the group-to-group scattering.

The following sections contain, as a matter of example, summarized explanations of the AZKIND nuclear reactor simulator in which the reactor physics is based on neutron diffusion theory.

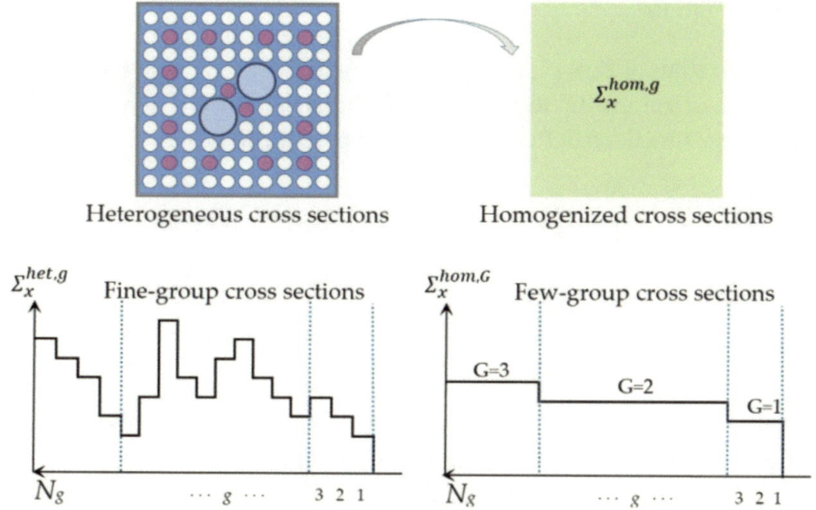

Figure 2. Homogenization and group collapsing of cross sections [2].

3. Neutron diffusion theory and nodal methods

3.1. Multi-group time-dependent neutron diffusion equations

For G neutron energy groups and I_p delayed neutron precursor concentrations, the neutron diffusion kinetics equations are given by Eqs. (1) and (2) [5]. Although there has been a growing interest in the transport-based core neutronics analysis methods for more accurate calculation with high-performance computers, it is yet impractical to apply them in the real core design activities because their performance is not so practical on ordinary desktop or server computing machines. For this reason, most of the neutronics codes for reactor core calculations are still subject to the two-step calculation procedure, which consists of homogenized group neutron parameter generation and neutron diffusion core calculation

$$\frac{1}{v^g}\frac{\partial}{\partial t}\phi^g\left(\vec{r},t\right) = \nabla \bullet D^g \nabla \phi^g\left(\vec{r},t\right) - \Sigma_R^g\left(\vec{r},t\right)\phi^g\left(\vec{r},t\right) + \sum_{\substack{g'=1 \\ g'\neq g}}^{G} \Sigma_s^{g'\to g}\left(\vec{r},t\right)\phi^{g'}\left(\vec{r},t\right)$$
$$+ (1-\beta)\chi^g \sum_{g'=1}^{G} \nu^{g'}\left(\vec{r},t\right)\Sigma_f^{g'}\left(\vec{r},t\right)\phi^{g'}\left(\vec{r},t\right) + \sum_{i=1}^{I_p} \chi_i^g \lambda_i C_i\left(\vec{r},t\right); \quad g = 1, \dots, G; \tag{1}$$

$$\frac{\partial}{\partial t}C_i\left(\vec{r},t\right) = \beta_i \sum_{g=1}^{G} \nu^g\left(\vec{r},t\right)\Sigma_f^g\left(\vec{r},t\right)\phi^g\left(\vec{r},t\right) - \lambda_i C_i\left(\vec{r},t\right); \quad i = 1, \dots, I_p; \quad \forall\left(\vec{r},t\right) \in \Omega \times (0,T] \tag{2}$$

In addition to boundary conditions for neutron fluxes, initial conditions must be satisfied by neutron fluxes and neutron precursor functions. Parameters involved in the above equations are described in [5].

3.2. Spatial discretization

The spatial discretization of Eqs. (1) and (2) is strongly connected with the discretization of a nuclear reactor core of volume Ω. Representing the neutron flux and the precursor concentrations in terms of base functions defined over Ω, it is possible to write

$$\phi^g\left(\vec{r},t\right) \equiv \sum_{k=1}^{N_f} u_k\left(\vec{r}\right)\phi_k^g(t);\ \ g = 1, \ldots, G;\ \ \forall\left(\vec{r},t\right) \in \Omega \times (0,T]; \tag{3}$$

$$C_i\left(\vec{r},t\right) \equiv \sum_{m=1}^{N_p} v_m\left(\vec{r}\right)C_i^m(t);\ \ i = 1, \ldots, I_p;\ \ \forall\left(\vec{r},t\right) \in \Omega \times (0,T]; \tag{4}$$

where N_f and N_p are the number of unknowns to be determined for neutron flux and delayed neutron precursors, respectively. Substituting expressions (3) and (4) into (1) and (2), and applying the Galerkin process for spatial discretization, as described in [6], the resulting algebraic system of equations can be expressed in a matrix notation as follows:

$$\begin{aligned}\frac{1}{v^g}\boldsymbol{M}_f\frac{d}{dt}\phi^g(t) = &-\boldsymbol{K}^g\phi^g(t) - \sum_{g'=1}^{G}\boldsymbol{S}^{g'\to g}\phi^{g'}(t) \\ &+ (1-\beta)\chi^g\sum_{g'=1}^{G}\boldsymbol{F}^{gg'}(t)\phi^{g'}(t) + \sum_{i=1}^{I_p}\boldsymbol{H}^{gi}(t)C_i(t),\ \ g = 1, \ldots, G;\end{aligned} \tag{5}$$

$$\boldsymbol{M}_p\frac{d}{dt}C_i(t) = \sum_{g'=1}^{G}\boldsymbol{P}^{ig'}(t)\phi^{g'}(t) - \lambda_i\boldsymbol{M}_pC_i(t),\ \ i = 1, \ldots, I_p;\ \ \forall\left(\vec{r},t\right) \in \Omega \times (0,T]; \tag{6}$$

where $\phi^g(t) = \left[\phi_1^g(t), \ldots, \phi_{N_f}^g(t)\right]^T$ and $C_i(t) = \left[C_i^1(t), \ldots, C_i^{N_p}(t)\right]^T$. **Table 1** contains the expressions representing the calculation of each matrix coefficient.

Matrix	Type	Dimension	Elements
$\boldsymbol{M}_f$	Mass	$N_f \times N_f$	$m_{f,jk} = \int_\Omega u_j u_k d\vec{r}$
$\boldsymbol{M}_p$	Mass	$N_p \times N_p$	$m_{p,lm} = \int_\Omega v_l v_m d\vec{r}$
$\boldsymbol{K}^g$	Stiffness	$N_f \times N_f$	$k_{jk}^g = \int_\Omega D^g \nabla u_j \cdot \nabla u_k d\vec{r}$
$\boldsymbol{S}^{g'\to g}$	Mass	$N_f \times N_f$	$s_{jk}^{g'\to g} = \int_\Omega \Sigma_s^{g'\to g} u_j u_k d\vec{r}$
$\boldsymbol{F}^{gg'}$	Mass	$N_f \times N_f$	$f_{jk}^{gg'} = \chi^g \int_\Omega \nu^{g'} \Sigma_f^{g'} u_j u_k d\vec{r}$
$\boldsymbol{H}^{gi}$	Mass	$N_f \times N_p$	$h_{jm}^{gi} = \lambda_i \chi^g \int_\Omega v_j u_m d\vec{r}$
$\boldsymbol{P}^{ig'}$	Mass	$N_p \times N_f$	$p_{lk}^{ig'} = \beta_i \int_\Omega \nu^{g'} \Sigma_f^{g'} u_l v_k d\vec{r}$

Table 1. Matrix elements from the spatial discretization.

http://dx.doi.org/10.5772/intechopen.79723

3.3. NFE method in spatial discretization

As fully explained in [6] and summarized in [1], a simple NFE element is characterized by the fact that for each node, the function unknowns to be determined are the (00) Legendre moment (average) of the unknown function over each face of the node and the (000) Legendre moment over the node volume. **Figure 3(a)** shows a physical domain Ω graphically represented after generating an *xyz* mesh. **Figure 3(b)** shows a cuboid-type node with directions through the faces: (*x*) Right, Left; (*y*) Near, Far; (*z*) Top, Bottom; and C for the average of the function over the node volume. Taking into consideration the general form to build up nodal schemes [7], the moments of a function (at edges and body) over a node like the one shown in **Figure 3(b)** can be written for the NFE method RTN-0 (Raviart-Thomas-Nédélec).

In the NFE method RTN-0, the normalized zero-order Legendre polynomials defined over the unit cell $\Omega_{ijk} = [-1,+1] \times [-1,+1] \times [-1,+1]$ and correlated to each physical cell $\Omega_e = \Omega_{ijk} = [x_i, x_{i+1}] \times [y_j, y_{j+1}] \times [z_k, z_{k+1}]$ are used to calculate the elements of the matrices in Eqs. (5) and (6).

The matrix elements are quantified introducing the following nodal basis functions [7]:

$$\begin{gathered} u_L^{00}(x,y,z) = -\frac{1}{2}(P_{100} - P_{200});\, u_R^{00}(x,y,z) = +\frac{1}{2}(P_{100} + P_{200}); \\ u_N^{00}(x,y,z) = -\frac{1}{2}(P_{010} - P_{020});\, u_F^{00}(x,y,z) = +\frac{1}{2}(P_{010} + P_{020}); \\ u_B^{00}(x,y,z) = -\frac{1}{2}(P_{001} - P_{002});\, u_T^{00}(x,y,z) = +\frac{1}{2}(P_{001} + P_{002}); \\ u_C^{000}(x,y,z) = P_{000} - P_{200} - P_{020} - P_{002}; \end{gathered} \tag{7}$$

where $P_{lpq}(x,y,z) = P_l(x)P_p(y)P_q(z)$.

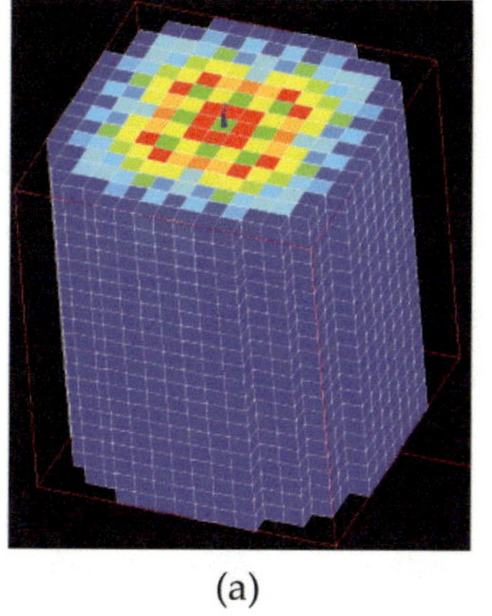

(a)

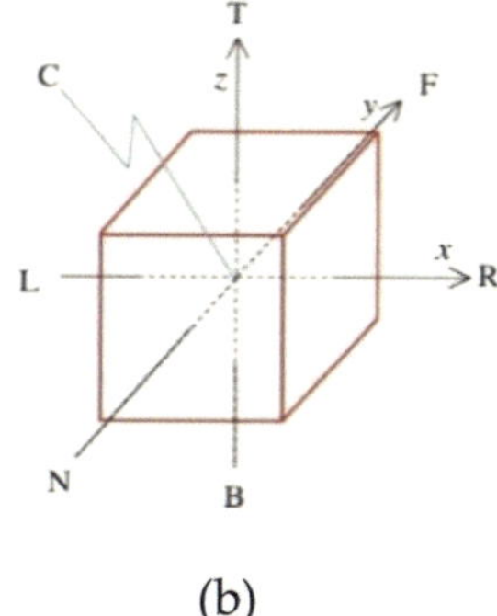

(b)

Figure 3. Discretization of reactor volume Ω and a local node Ω_e. (a) Domain Ω. (b) Physical local node Ω_e.

An extensive discussion on nodal diffusion methods can be found in Ref. [7] for space discretization using simplification approaches for calculating the moments over a node.

3.4. Discretization of the time variable

Once the spatial discretization is done, the θ-method can be applied [6] for the discretization of the time variable appearing in the algebraic system given by (5) and (6). For the time integration over the interval $(0, T]$, this interval is divided in L time-steps $[t_l, t_{l+1}]$, and the following approach is assumed:

$$\int_{t_l}^{t_{l+1}} f(t)dt \cong h_l\left[\theta f_{l+1} + (1-\theta)f_l\right] \tag{8}$$

where $h_l = t_{l+1} - t_l, f_l = f(t_l), f_{l+1} = f(t_{l+1})$, and θ is the time integration parameter.

For time integration, parameters θ_f and θ_p for neutron flux and delayed neutron precursors are considered with values in the interval [0, 1], giving different time integration schemes [6].

Once the formulation to be used for time integration is established, the $N_fG + N_pI$ system of equations that was spatially discretized, Eqs. (5) and (6) are discretized over the interval $(0,T]$. Integrating the referred equations over the time interval $[t_l, t_{l+1}]$ using approximation (8), the following set of equations is generated:

$$\boldsymbol{A}_{l+1}\Phi_{l+1} = Q_l; \quad l = 0, 1, 2, \ldots,; \tag{9}$$

$$\Phi_{l+1} = \left[\phi_{l+1}^1, \ldots, \phi_{l+1}^G\right]^T; \quad Q_l = \left[S_{f,l}^1, \ldots, S_{f,l}^G\right]^T;$$

For a known vector Φ_l the algebraic system (9) is solved for the neutron fluxes Φ_{l+1}. Therefore, the computing process requires an initial flux vector for the first time step, which is used in (9) to determine new neutron fluxes at the end of the time step, thus using these neutron fluxes to calculate a new delayed neutron precursor concentration vector. This process is sequentially performed for each time step over the total time interval $(0,T]$.

4. Reactor power distribution

Once the computer model to solve the reactor kinetics Eqs. (1) and (2) is able to provide the neutron flux profile, the next objective is to know the power distribution in the reactor configuration. It is necessary to be aware that the neutron flux is by itself the shape of the power distribution in multiplicative materials. The numerical methods presented in previous sections to solve Eq. (9) produce an algorithm capable to obtain the neutron flux profile for a reactor steady state. The calculated neutron flux has the following property over the domain Ω: $\|\phi\| = 1$. To determine the real average neutron flux in the reactor core, ϕ_c, it is necessary to specify the magnitude of the fluxes. For instance, a flux normalization factor ϕ_{norm} can be introduced such that $\phi_c = \phi_{norm}\phi\left[\frac{neutrons}{cm^2 \cdot seg}\right]$.

http://dx.doi.org/10.5772/intechopen.79723

Theoretically, it would be best to determine the flux level which resulted in a critical reactor (*eigenvalue* $\lambda_0 = 1$). This could be accomplished by coupling of the NK model with the TH model of the whole reactor. In practice, however, the scaling factor ϕ_{norm} is determined such that the total generated thermal power corresponds to some user-specified value $P_{th,tot}$. Before showing how this is done, the relation between the fluxes and the generated thermal power is described. For a given discretization of the xy-plane with pieces of area $\Delta a = \Delta x \cdot \Delta y$, the thermal power $P_{th,tot}$ can be expressed as follows:

$$P_{th,tot} = \sum_{\Delta a} \int_{z_b}^{z_t} q_f'''(z) da \cdot dz, \quad dV = da \cdot dz; \tag{10}$$

where q_f''' is the volumetric heat generation rate in the fuel in units of [W/cm^3], dV is a differential fuel volume, and the limits z_b and z_t refer to the coordinates of the bottom and top of the reactor core, respectively. For a given area Δa, the volumetric heat generation rate $q_f'''(z)$ in an elevation z may be written in terms of the fluxes as

$$q_f'''(z) = \phi_{norm} E_{fiss} \sum_{g'=1}^{G} \Sigma_f^{g'}(z) \phi^{g'}(z); \tag{11}$$

where ϕ_{norm} is a dimensionless factor, E_{fiss} is the energy released by a nuclear fission reaction in [MeV/fission], and the sum over g′ is the volumetric fission rate in [fissions/(cm^3·s)]. Thus, Eq. (10) is written as

$$P_{th,tot} = \phi_{norm} E_{fiss} \sum_{\Delta a} \int_{z_b}^{z_t} \sum_{g'=1}^{G} \Sigma_f^{g'}(z) \phi^{g'}(z) da \cdot dz. \tag{12}$$

In a more general way, for a reactor volume V composed by the union of sub-volumes V_e (see **Figure 3**), the total thermal power can be expressed as

$$P_{th,tot} = \phi_{norm} E_{fiss} \sum_{e=1}^{N_e} \sum_{g'=1}^{G} \Sigma_{f,e}^{g'} \phi_e^{g'} V_e. \tag{13}$$

Therefore, using the reference total thermal power specified by the code user, the flux normalization factor can be written as

$$\phi_{norm} = P_{th,tot} \left[\sum_{e=1}^{N_e} \sum_{g'=1}^{G} \kappa_{f,e}^{g'} \phi_e^{g'} V_e \right]^{-1}; \tag{14}$$

where the factors "kappa-fission" are $\kappa_{f,e}^{g'} = E_{fiss} \Sigma_{f,e}^{g'}$. With the flux normalization factor ϕ_{norm} calculated as above, the actual thermal power distributions in the reactor core can be calculated using the current neutron flux in the reactor core $\phi_c^e = \phi_{norm} \phi^e$. Nevertheless, it is necessary to introduce the value of E_{fiss}. This value is used as an average energy released of ~200 MeV (i.e.,), based on the energies released by the fission of the U^{235} nuclei [8].

In summary, once the NK model is used to generate the neutron flux distribution in the reactor core, expression (12) can be used to calculate the thermal power being generated along all the nodes in a thermal-hydraulic channel of area Δa and height H. This thermal power can be the axial power profile needed by the TH model to produce the thermal-hydraulic state corresponding to the generated thermal power.

5. Neutronic and thermal-hydraulic coupling model (NK-TH)

The description contained in this section is based on a work published by Ceceñas in Ref. [9] about a TH model developed for boiling water reactors. The TH model was modified from a point kinetics approach with an extension of the NK model to 3D and implemented in the development of AZKIND.

The treatment of neutron kinetics in [9] has been improved by coupling a 3D solution of the neutron diffusion equations with an arrangement of TH channels in parallel. Each channel independently contemplates three regions: (1) one phase, (2) subcooled boiling, and (3) bulk boiling. The objective was to implement a detailed model of a nuclear reactor core, which is somehow perturbed to simulate NK-TH coupling. These perturbations are obtained when the power generated in a group of channels changes and thus affecting the TH state of each channel.

The original [9] TH model is based on a generic channel, which is adapted by transferring to it the operational data as flow area, generated power, axial power profile, and subcooling, among other parameters. Each channel is associated with a number of nuclear fuel assemblies and an axial power profile. Although the neutron model is a two-dimensional model for the radial power profile in each z-plane covering all the channels, information related to the axial power distribution is considered for each individual channel. In Ref. [9], it is assumed that this steady-state axial power profile is invariant over time, and it is used to weight the axial averages of macroscopic cross sections and void fractions. To perform the numerical implementation of the model, the arrangement of channels is obtained by grouping the total core assemblies into an appropriate number of thermal-hydraulic channels, which gives a definition of a set of channels per quadrant.

For the implementation in AZKIND of the TH model of Ceceñas, the grouping of fuel assemblies was maintained for generating a reduced number of TH channels; operational data are also used. The main difference is that the NK model recursively computes the axial power profile for each channel, and this thermal power is the updated source of power for TH model. Therefore, a "new" thermal-hydraulic condition is generated, and it is used by the NEMTAB model to update the nuclear data to generate new thermal power profiles with the NK model. The process is iterative, and it stops when the convergence is met. Convergence is achieved when updated conditions do not change in both NK and TH models.

The NK-TH coupling in AZKIND performs core calculations as described above to obtain a steady-state reactor core condition. For transient conditions in a time interval T, the NK-TH coupling process is the same for each time step ΔT in T, that is, a different quasi-steady-state condition for each successive ΔT. Achieving converge for each ΔT with respective reactor core conditions means to produce a time-dependent behavior of the reactor condition over the total time interval T.

The TH model comprises the solution of the mass, momentum, and energy conservation equations in the three regions contemplated by the channel: (1) one phase, (2) subcooled boiling, and (3) bulk boiling. The system receives heat through a non-uniform source whose profile is axially defined plane by plane. This axial use of the power profile allows the inclusion of a wide range of axial profiles, from relatively flat to profiles with their peak value at some axial point in each channel in the core.

In the following subsections, there are several expressions for which the corresponding parameters are defined in Refs. [10, 11].

5.1. Heat transfer in the fuel

The heat transfer and temperature distribution in the fuel and cladding can be calculated by a simple model where the heat diffusion equation is solved in one dimension (radial) for a fuel rod, since the conduction in axial direction is small compared to the radial one, it can be neglected. An energy balance per unit length yields

$$m_f c_{pf} \frac{dT_f}{dt} = q'(t) - \frac{1}{R'_g}\left[T_f(t) - T_c(t)\right] \tag{15}$$

$$m_c c_{pc} \frac{d\overline{T_c}}{dt} = \frac{1}{R'_g}\left[\overline{T_f}(t) - \overline{T_c}(t)\right] - \frac{1}{R'_c}\left[\overline{T_c}(t) - \overline{T_m}(t)\right] \tag{16}$$

where R'_g and R'_c represent thermal resistances per unit length. The coefficient of heat transfer to the refrigerant fluid is calculated by the Dittus-Boelter or Chen correlation, depending on the type of flow, which can be in one or two phases. These equations are used for the radial averaging of the temperatures in the fuel rod.

5.2. Reactor coolant dynamics

The conservation equations of mass, energy, and momentum are applied in this case to a flow of water along a vertical channel, where the dynamics of the fluid heated by the wall of the fuel is modeled. Conservation equations can be expressed as [10]

$$\frac{\partial \rho_m}{\partial t} + \frac{\partial G_m}{\partial z} = 0 \tag{17}$$

$$\frac{\partial G_m}{\partial t} + \frac{\partial}{\partial z}\left(\frac{G_m^2}{\rho_m^+}\right) = -\frac{\partial p}{\partial z} - \frac{fG_m|G_m|}{2D_e\rho_m} - \rho_m g\cos\theta \tag{18}$$

$$\rho_m \frac{\partial h_m}{\partial t} + G_m \frac{\partial h_m}{\partial z} = \frac{q'' P_h}{A_z} + \frac{\partial p}{\partial t} + \frac{G_m}{\rho_m}\left(\frac{\partial p}{\partial z} + \frac{fG_m|G_m|}{2D_e\rho_m}\right) \tag{19}$$

In this work, the conservation equations are solved by the Integral Moment method [11], according to which it is assumed that the refrigerant is incompressible but thermally expandable, and the density is a function of enthalpy at a constant pressure

$$\frac{\partial \rho_m}{\partial t} = \left.\frac{\partial \rho_m}{\partial h_m}\right|_p \frac{\partial h_m}{\partial t} + \left.\frac{\partial \rho_m}{\partial p}\right|_{h_m} \frac{\partial p}{\partial t} = R_h \frac{\partial h_m}{\partial t} + R_p \frac{\partial p}{\partial t} \tag{20}$$

Neglecting terms related to pressure changes and wall friction forces, the energy equation is simplified as

$$\rho_m \frac{\partial h_m}{\partial t} + G_m \frac{\partial h_m}{\partial z} = \frac{q'' P_h}{A_z} \tag{21}$$

where the axial flow variation can be obtained by

$$\frac{\partial G_m}{\partial z} = -\frac{R_h}{\rho_m}\left(\frac{q'' P_h}{A_z} - G_m \frac{\partial h_m}{\partial z}\right) \tag{22}$$

This equation provides the flow variations with respect to an average value imposed as a boundary value or provided by the dynamics of the coolant recirculation system. Three regions are defined by which the coolant circulates as it ascends into the channel: a one-phase region, a subcooled boiling region, and a bulk boiling region. The first region begins at the bottom of the channel, where the coolant enters with known enthalpy and ends at the point of separation of the bubbles Z_{sc}. The bulk temperature at this point is obtained by the Saha and Zuber correlation. The subcooled boiling region ends when the bulk temperature reaches the saturation temperature, and its axial location is determined by an energy balance. The enthalpy distribution allows the calculation of the thermodynamic equilibrium quality, used to calculate the flow quality. The axial distribution of the void fractions is calculated by iteratively solving the equation for void fraction α and the Bankoff correlation slip (S):

$$\alpha = \frac{x}{S\left(\frac{\rho_g}{\rho_f}\right) + x\left(1 - S\left(\frac{\rho_g}{\rho_f}\right)\right)}, \; S = \frac{1-\alpha}{k_s - \alpha + (1-k_s)\alpha^r}, \tag{23}$$

where, in this case, the parameters k_s and r are functions of the system pressure:

$k_s = 0.71 + 1.2865 \times 10^{-3}p$, and, $r = 3.33 - 2.56021 \times 10^{-3}p + 9.306 \times 10^{-5}p^2$.

The total pressure drop in the channel is made up of the contributions of each region. Every term in each region includes the contribution by acceleration, gravity, and friction. For the channel arrangement, the steady state is obtained by iterating the coolant flow rate of each channel to obtain the same pressure drop for all of them. This iteration consists of a correction to the flow defined by the deviation of the pressure drop of the channel with respect to the average of all the channels:

$$G_i^{k+1} = G_i^k + w G_i^k \left(\frac{\overline{P}^k - P_i^k}{P_i^k}\right) \tag{24}$$

where G_i is the flow rate for channel i, the index k represents the number of the iteration, w is an arbitrary weight to control the convergence, and P is the average pressure drop of all channels at iteration k, obtained as

$$\overline{P}^k = \frac{1}{N}\sum_{1=1}^{N} P_i^k \tag{25}$$

It is observed that even though the pressures are equaled, the value of the pressure drop in the core is not imposed as a boundary condition. Convergence is achieved when the following relationship is met: $\sum_{i=1}^{N} \left|\overline{P}^k - P_i^k\right| < \varepsilon$. By changing the flow rate of the channel for each iteration, the enthalpy and void fraction profiles are affected. It is necessary to recalculate the TH solution at each iteration for all channels, achieving convergence when every parameter involved in the thermal-hydraulic calculation remains unchanged.

5.3. Neutron kinetics: thermal-hydraulics (NK-TH) coupling model

Although reference [12] has important issues to be considered in the development of an NK-TH-coupled model, those issues are not repeated here, but taken into account. The most direct way of coupling NK module and TH module, as implemented in AZKIND, consists simply in that axially both NK mesh and TH mesh have the same partition, making possible to assign an NK node at position z to the TH node in the same position. This relationship is a one-to-one node correspondence.

As it can be seen in **Figure 4**, before initiating the NK-TH feedback process, the initial nuclear parameters and kinetics parameter (XS) are loaded from files constructed in NEMTAB format, previously generated by means of a lattice code. Then, following the reading of the nuclear reactor burn-up state and thermo-physics initial conditions, the XS parameters are obtained from the Nemtab multi-dimensional tables by means of interpolation calculations.

The process continues as follows. The corresponding neutron flux is calculated in the NK module with the *mgcs* numerical solver, and this power (initial neutron flux) is the heat source to be assigned to the TH model. The axial power profile can be that of each fuel assembly assigned to a unique TH channel or the power profile of a set of fuel assemblies assigned to a TH channel. The axial power profile is the heat source for each node in the z-direction. Once the axial power profiles have been constructed in the TH module, an initial thermal-hydraulic state of the reactor system is calculated. The thermal-hydraulic state is calculated for each node in the TH channels from the bottom to the top of the reactor core.

The important variables sent to the NK module are the fuel temperature (T_f), moderator temperature (T_m), and moderator density (*Dens*). The XS parameters are updated using these 3D variables for interpolation in the NEMTAB tables. The next step is to calculate new 3D power profiles to be sent to the TH module. This cyclic NK-TH calculation continues and stops when the TH criterion and neutron-flux criterion are met. Stopping the cyclic calculation means that the reactor power and thermal-hydraulic conditions have reached a steady state.

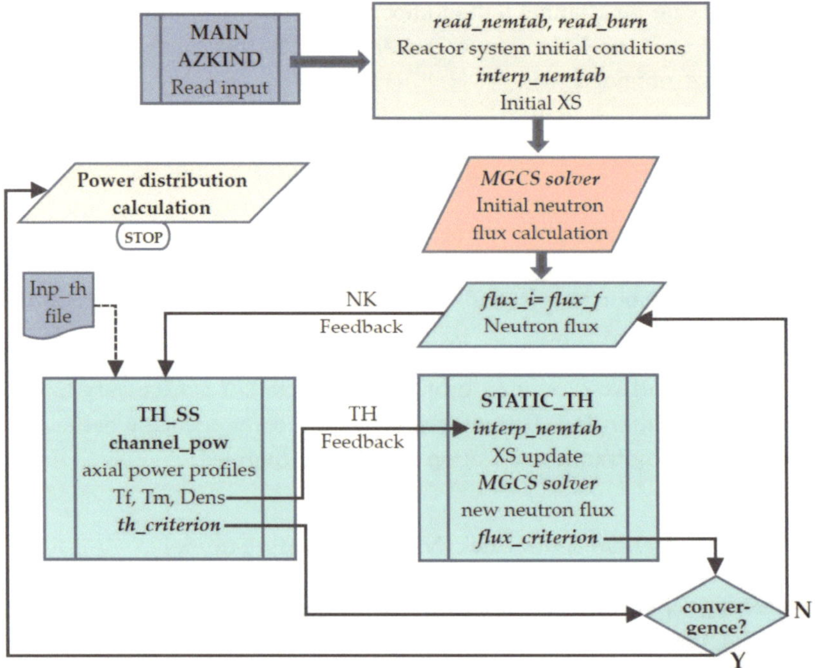

Figure 4. The NK-TH feedback process in AZKIND.

6. High-performance computing in AZKIND

6.1. PARALUTION linear algebra library

HPC was implemented in AZKIND with the support of the linear algebra solvers library PARALUTION [13]. This open-source library is optimized for parallel computing process using graphics processing units (GPUs). For the numerical solution of an algebraic system $A\ \vec{v}=\vec{b}$ PARALUTION includes numerical solvers to obtain the solution vector $\vec{v}$ for a known vector $\vec{b}$ and a specific matrix A that can be a symmetric or a non-symmetric matrix being also a sparse or a dense matrix. The working matrices in AZKIND are sparse non-symmetric matrices, and the *bicgstab* solver [14] was used for reactor simulations. The matrix solvers in PARALUTION are optimized to use on the non-zero (*nnz*) elements in the working matrices, saving processing time and computer memory.

6.2. Parallel processing for neutronic model

To demonstrate the HPC implementation in AZKIND, as described in Ref. [1], very large matrices were constructed for fine spatial discretization of arrangements of nuclear fuel assemblies of an LWR. Fine discretization means that each fuel assembly was subdivided in a mesh of size 10 × 10. As an example, an arrangement of 6 × 6 fuel assemblies consists of a square with 36 fine-discretized fuel assemblies. The corresponding algebraic system for each fuel arrangement was solved with parallel processing performed by the *bicgstab* solver mentioned earlier. In **Tables 2** and **3**, the speedup of the different cases is shown [1] with a remarkable performance. Despite the speedup for small matrices that is comparable for the three computer architectures used, it is also important to notice that the speedup values listed in **Table 3** do not present a linear behavior, and the reason is because although more GPU processor cores are used with massive data transference to and from the GPU, a data traffic delay is present in the communication bus between the GPU and the CPU. For the analysis of the computing acceleration or "speedup," a definition of speedup is used in [15], known as relative speedup or speedup ratio: $S = T_1/T_n$, where T_1 is the computing time using a single processor (serial calculation) and T_n is the computing time using n processor cores. The "no memory" insert listed in **Table 2** is because for those large matrix dimensions, there is not enough memory to load the matrix and solvers.

Figure 5 [1] shows the distribution of nuclear fuel assemblies in the core of a boiling water reactor. Excepting the blue-shaded zone, colors are for different types of fuel assemblies. In the plane *xy*, the mesh is 24 × 24, according to each fuel zone, and axially, there are 25 nodes. The matrix for this coarse mesh (1,274,304 *nnz*) is comparable to the matrix of the fine mesh created for the case of a unique assembly (case 1 × 1 listed in **Table 2**).

As described in [1], a reactor power transient was simulated as the capability to remove neutrons was highly increased in the perturbed assembly shown in **Figure 5**. An increase as

Assemblies array: **Matrix dimension (*n*):** ***nnz* elements:**	**1 × 1** **126,200** **1,332,400**	**2 × 2** **492,800** **5,305,600**	**4 × 4** **1,947,200** **21,174,400**	**6 × 6** **4,363,200** **47,606,400**	**10 × 10** **12,080,000** **132,160,000**
Serial	24	124	372	994	2471
GTX 860 M	2.1	7.9	31.3	No memory	No memory
Tesla K20c	1.3	4.0	16.6	40.1	No memory
GTX TITAN X	1.0	2.6	10.4	26.7	95.4

Table 2. Parallel processing time (seconds) in different architectures [1].

Assemblies	**1 × 1**	**2 × 2**	**4 × 4**	**6 × 6**	**10 × 10**
GTX 860 M	11	16	12	–	–
Tesla K20c	18	31	22	24	–
GTX TITAN X	24	48	36	37	26

Table 3. Speedup comparison (S) [1].

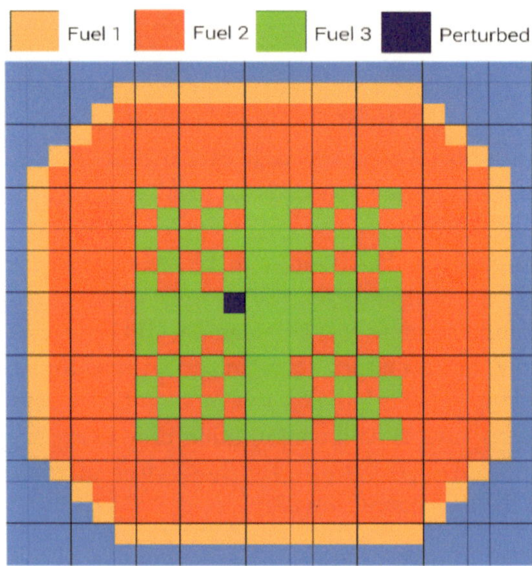

Figure 5. A map of fuel assemblies in an LWR [1].

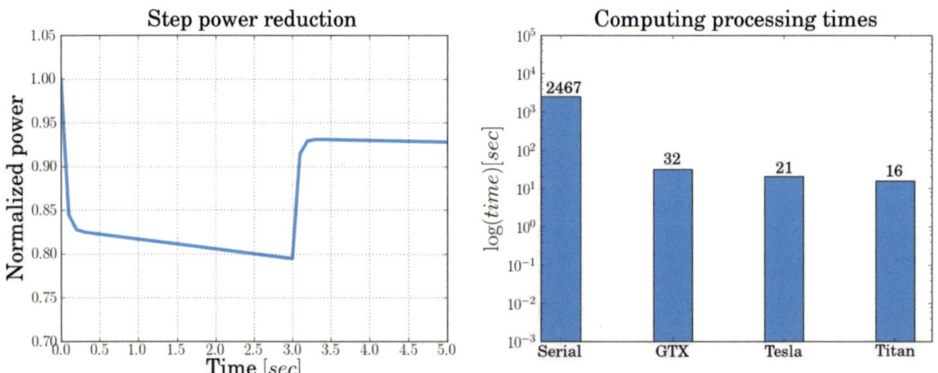

Figure 6. Simulation of a reactor power transient—serial and parallel processing.

step function in the neutrons removal capability during 3 s is implemented in the perturbed assembly, after that the perturbation finishes and the transient lasts for two more seconds, giving a reactor power reduction. The time step used in this simulation was 0.1 s. **Figure 6** shows the power behavior over time, departing from a normalized value of 1.0 and reducing the power reactor to almost 80% of its original value. This reactor power transient was simulated with the AZKIND code, running on the three different GPUs listed in **Tables 2** and **3**. The right side of **Figure 6** shows the time spent by AZKIND in a logarithmic scale, running in a sequential mode (Serial bar) and the times spent by each GPU card.

http://dx.doi.org/10.5772/intechopen.79723

7. Simulation of a reactor core condition

A simple example was prepared to show the capability of the AZKIND code running with NK-TH coupling, and the thermal-hydraulic effect on power distribution is compared to the power distribution resulted from the NK model running standalone.

This example was prepared for a two energy group, that is, fast neutrons and thermal neutrons. In LWR, the nuclear fissions of the fuel atoms are mainly coming from the thermal neutrons present in the reactor core. The effect observed in **Figure 7** is that the TH feedback induces an increase in the thermal neutrons population and so increasing power. As the coolant/moderator enters the reactor core through the bottom part of the reactor and the core

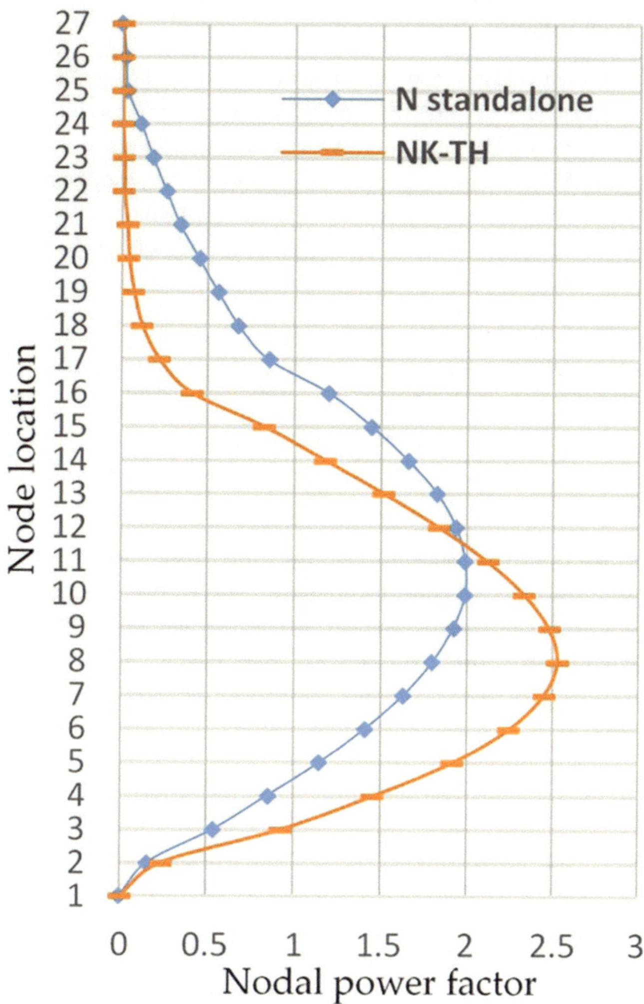

Figure 7. Axial power peaking profile location.

is beginning the production cycle, the core design allows more power generation in the first third of the core active fuel. Also, as it was expected, in the map of fuel assemblies of the reactor core, the location of the fuel assembly with the highest generation of thermal power remained unchanged with the insertion of TH feedback.

8. Some advances on nuclear reactor simulation

In the last two decades, there have been significant advances in the development of nuclear reactor codes for 3D simulation with coupling NK-TH, supported with new modeling techniques and modern computing capabilities in software and hardware. Some examples of these advances are listed subsequently:

1. DYNSUB: Pin-based coupling of the simplified transport (SP3) version of DYN3D with the sub-channel code SUBCHANFLOW. See [16, 17]. The new coupled code system allows for a more realistic description of the core behavior under steady state and transient conditions. DYNSUB has successfully been applied to analyze the behavior of one eight of a PWR core during an REA transient by a pin-by-pin simulation consisting of a huge number of nodes. Some insights are pointed out on the convergence process with a detailed coupling solution modeling neighbor sub-channels and modeling adjacent assembly channels.

2. DYN3D: The code comprises various 3D neutron kinetics solvers, a thermal-hydraulics reactor core model, and a thermo-mechanical fuel rod model, see [18]. The following topics are delineated in the reference: the latest developments of models and methods, a status of verification and validation; code applications for selected safety analyses; multi-physics code couplings to thermal-hydraulic system codes, CFD, and sub-channel codes as well as to the fuel performance code TRANSURANUS.

3. TRACE/PARCS: See [19]. The study of the coupling capability of the TRACE and PARCS codes by analyzing the "Main Steam Line Break (MSLB) benchmark problem," consisting of a double-ended MSLB accident assumed to occur in the Babcock and Wilcox Three Mile Island Unit 1. The model TRACE/PARCS generated data showing that these codes have the capability to predict expected phenomena typical of this transient and the related NK-TH feedback.

4. COBAYA3: See [20]. This reference describes a multi-physics system of codes including the 3D multi-group neutron diffusion codes, ANDES and COBAYA3-PBP, coupled with the sub-channel thermal-hydraulic codes COBRA-TF, COBRA-IIIc, and SUBCHANFLOW, for the simulation of LWR core transients. Implementation of the PARALUTION library to solve sparse systems of linear equations was done. It features several types of iterative solvers and preconditioners which can run on both multi-core CPUs and GPU devices without any modification from the interface point of view. By exploring this technology, namely the implementation of the PARALUTION library in COBAYA3, the code can decrease the solution time of the sparse linear systems by a factor of 5.15 on GPU and 2.56 on a multi-core CPU using standard hardware.

5. CNFR: See [21]. This reference summarizes three methods, implemented for multi-core CPU and GPU, to evaluate fuel burn-up in a pressurized light water nuclear reactor (PWR) using the solutions of a large system of coupled ordinary differential equations. The reactor physics simulation of a PWR with burn-up calculations spends long execution times, so that performance improvement using GPU can imply in a better core design and thus extended fuel life cycle. The results with parallel computing exhibit speed improvement exceeding 200 times over the sequential solver, within 1% accuracy.

9. Conclusions and remarks

The state of the art in the topic of nuclear reactor simulations shows significant advances in the development of computer codes. A wide range of applications focusing, besides on improving nuclear safety, on more efficient analyses to improve fuel cycles/depletion have been found in a recent study. A considerable "saving time" factor in obtaining nuclear reactor analyses has been observed.

One important part of a nuclear reactor simulator is the benchmarking process to demonstrate reliability and repeatability in the simulation of real cases, for which data from reactor operation or comprehensive data from experiments are well documented. In this sense, extensive documentation is necessary for theoretical basis, numerical techniques and tools, and validation of both codes and simulation models.

Author details

Andrés Rodríguez Hernández[1], Armando Miguel Gómez-Torres[1]* and Edmundo del Valle-Gallegos[2]

*Address all correspondence to: armando.gomez@inin.gob.mx

1 Instituto Nacional de Investigaciones Nucleares, Ocoyoacac, Edo. de México, México

2 Instituto Politécnico Nacional, Escuela Superior de Física y Matemáticas, Col. San Pedro Zacatenco, Cd. de México, México

References

[1] Rodríguez-Hernández A, Gomez-Torres A, Del Valle-Gallegos E. HPC implementation in the time-dependent neutron diffusion code AZKIND. Annals of Nuclear Energy. 2017;**99**: 174-182

[2] Oka Y, editor. Nuclear Reactor Design. (Series) An Advanced Course in Nuclear Engineering. Japan: Springer; 2014

[3] Fridman E, Leppänen J. Revised methods for few-group cross sections generation in the Serpent Monte Carlo code. PHYSOR 2012 – Advances in Reactor Physics. LaGrange Park, IL: American Nuclear Society; 2010

[4] Reedmond EL. Multigroup cross section generation via Monte Carlo methods [PhD thesis]. Massachusetts Institute of Technology; 1998

[5] Duderstadt JJ, Hamilton LJ. Nuclear Reactor Analysis. New York: John Wiley and Sons; 1976

[6] Rodríguez-Hernández A. Solution of the nuclear reactor kinetics equations in 3D using the nodal method RTN-0 (in Spanish) [MSc thesis]. México: National Polytechnic Institute, ESFM; 2002

[7] Grossman LM, Hennart JP. Nodal diffusion methods for space-time neutron kinetics. Progress in Nuclear Energy. 2007;**49**:181-216

[8] Weismann J, editor. Elements of Nuclear Reactor Design. Elsevier Scientific Publishing Company; 1977

[9] Ceceñas M, Campos R. Modelo acoplado de canales en paralelo y cinética neutrónica en dos dimensiones. In: International Joint Meeting Cancun 2004 LAS/ANS-SNM-SMSR. Cancun, Mexico; 2004

[10] Todreas NE, Kazimi MS. Nuclear Systems I: Thermal Hydraulic Fundamentals. USA: Hemisphere Publishing; 1989

[11] Todreas NE, Kazimi MS. Nuclear Systems II: Elements of Thermal Hydraulic Design. USA: Hemisphere Publishing; 1990

[12] Ivanov K, Avramova M. Challenges in coupled thermal–hydraulics and neutronics simulations for LWR safety analysis. Annals of Nuclear Energy. 2007;**34**:501-513

[13] Lukarski D. PARALUTION Project, Version 0.8.0. 2014. http://www.paralution.com/

[14] Van der Vorst HA. Efficient and reliable iterative methods for linear systems. J. of Computational and Applied Mathematics. 2002;**149**:251-265

[15] Nesmachnow S. Workshop Scientific computing on distributed memory systems. In: International Supercomputing Conference ISUM; Mexico; 2015

[16] Gomez-Torres AM, Sanchez-Espinoza VH, Ivanov K, Macian-Juan R. DYNSUB: A high fidelity coupled code system for the evaluation of local safety parameters—Part I: Development, implementation and verification. Annals of Nuclear Energy. 2012;**48**:108-122

[17] Gomez-Torres AM, Sanchez-Espinoza VH, Ivanov K, Macian-Juan R. DYNSUB: A high fidelity coupled code system for the evaluation of local safety parameters—Part II: Comparison of different temporal schemes. Annals of Nuclear Energy. 2012;**48**:123-129

[18] Rohde U, Kliem S, Baier S, Bilodid Y, Duerigen S, Fridman E, Gommlich A, Grahn A, Holt L, Kozmenkov Y, Mittag S. The reactor dynamics code DYN3D e models, validation and applications. Progress in Nuclear Energy. 2016;**89**:170-190

[19] Mascari F, Vella G, Casamassima V, Parozzi F. Analyses of TRACE-PARCS coupling capability. In: International Conference on the Physics of Reactors. 20th International Conference—Nuclear Energy for New Europe; Slovenia; 2011

[20] Trost N, Jimenez J, Lukarski D, Sanchez V. Accelerating COBAYA3 on multi-core CPU and GPU systems using PARALUTION. Annals of Nuclear Energy. 2014;**82**:252-259

[21] Heimlich A, Silva FC, Martinez AS. Parallel GPU implementation of PWR reactor burnup. Annals of Nuclear Energy. 2016;**91**:135-141

Chapter 3

Mechanisms of Charge Transport and Photoelectric Conversion in CdTe-Based X- and Gamma-Ray Detectors

Olena Maslyanchuk, Stepan Melnychuk,
Volodymyr Gnatyuk and Toru Aoki

Additional information is available at the end of the chapter

http://dx.doi.org/10.5772/intechopen.78504

Abstract

This chapter deals with (i) the charge transport mechanisms in X- and gamma-ray detectors both Ohmic and Schottky types based on CdTe and its alloys with an almost intrinsic conductivity (the peculiarities of the formation of self-compensated complexes due to the doping of Cd(Zn)Te crystals with elements of III or V groups (In, Cl) are taken into account); (ii) the reasons of insufficient energy resolution in the X- and gamma-ray spectra taken with the detectors under study; (iii) the quantitative model which describes the spectral distribution of the detection efficiency of Cd(Zn)Te crystals with Schottky diodes; (iv) a correlation between the concentration of uncompensated impurities in the Cd(Zn)Te crystals and collection efficiency of photogenerated charge carriers in the detectors with a Schottky contact; (v) the possibility of applications of CdTe thin films with a Schottky contact as an alternative to the existing X-rays image detectors based on a-Se.

Keywords: X- and gamma-ray detector, CdTe, CdZnTe, CdMnTe, self-compensation, Schottky diode, concentration of uncompensated impurities, detection and collection efficiency

1. Introduction

The current state and development of technology, science, medicine and other fields of human activity are impossible without elemental analysis—a combination of methods of detection and quantitative determination of the elemental composition of objects of the material

world. The rapid development of the methods of elemental analysis began in the 1950s, when ionization chambers and scintillation devices were replaced with solid-state semiconductor devices—detectors of X- and γ-rays (X/γ-rays). Germanium (Ge) and silicon (Si) were the first materials for semiconductor detectors. Such detectors have high-energy resolution; however, cryogenic cooling is required to reduce "dark" electrical noise, which in many cases is impractical or even impossible. Because of the small atomic number Z, the absorptive capacity of Si (Z = 14) is low. Therefore, the registration of quanta with energies above 30–50 keV is practically impossible. The atomic number is Ge higher (Z = 32), but due to the narrower band gap, the problem of too large dark current becomes even more serious. For a long time, intensive search of alternative semiconductors for X/γ-rays detectors is being carried out. The main purpose is to reduce or even eliminate the disadvantages of Si- and Ge-detectors, namely, reducing of the "dark" current due to the expansion of the bandgap, detecting of higher energy quanta (>30–50 keV) due to the increase in Z and material density, and improving of the energy and time resolution due to increase in the life time and the mobility of charge carriers ($\tau\mu$ product). The possibilities of application as elementary semiconductors and insulators (e.g., diamond) as binary compounds (GaAs, GaP, HgI_2, PbI_2, CdSe, SiC, etc.) are studied. Currently, the most common semiconductors among binary and ternary II–VI compounds are CdTe and $Cd_{1x}Zn_xTe$.

Over 30 years, the heightened interest in CdTe and $Cd_{1-x}Zn_xTe$ detectors resulting from their high power and coordinate energy resolution, which allowed to expand their applications, in particular, in the devices for customs control and movement of dangerous goods, security systems in the airports, railway stations, transport highways, in crowded places [1–3]. Such detectors are also used in metallurgical, chemical, mining and nuclear industries, for radiation control by environmental services, for astrophysical applications. In addition, it is extremely promising to use CdTe as the base material in multielement (pixelated) detectors, which opens up the possibility of creating image detectors with direct X-rays conversion into electrical signals, in contrast to the "classical" X-ray devices where the detector is a screen, covered with a luminophore and hidden X-image can either observe or fix with a photoemulsion. A multielement detector allows you to convert information generated by passing X/γ-rays through an object directly into an array of digital electrical signals without using an intermediate visible image on a fluorescence screen. Such detectors allow real-time visualization of the X-ray image and sufficiently enhance the image resolution. The X- and γ-image CdTe detectors are already widely used in the flaw detection of materials with high spatial resolution, in medical tomographs with a small dose of a patient's irradiation, in mammographs, dental appliances, in the diagnosis of cancerous tumors.

Modern technology provides the growth of perfect CdTe crystals with almost intrinsic electrical conductivity. However, Ohmic detector's dark currents restrict the high-bias voltage for creation of a strong electric field which is necessary to full charge collection. In the 1990s, research of CdTe-based ternary compounds for X/γ-rays detectors has started. In particular, $Cd_{1-x}Zn_xTe$ (x = 0.1–0.2) crystals with a wider band gap, higher resistivity and lower dark currents in comparison with the CdTe detector has been studied. However, the expected prospects of $Cd_{1-x}Zn_xTe$ crystals remain unfulfilled because of high temperature of crystal growth, their crystalline imperfection, and the effect of segregation. This lack is absent in

$Cd_{1-x}Mn_xTe$ crystals, since the growth temperature of $Cd_{1-x}Mn_xTe$ crystals is lower than that of $Cd_{1-x}Zn_xTe$. Moreover, about two times less manganese in $Cd_{1-x}Mn_xTe$ than that zinc in $Cd_{1-x}Zn_xTe$ should be introduced for the expansion of the CdTe band gap. In the late 1990s, the first studies of the $Cd_{1-x}Mn_xTe$-based X/γ-ray detectors were carried out in the USA, Japan, and Europe. However, the spectrometric characteristics of the detectors are still significantly worse than those of CdTe and $Cd_{1-x}Zn_xTe$. At the end of the 1990s, CdTe detectors with Schottky diodes were developed. The detectors showed significantly better energy resolution in the wide range of photon energy up to 1 MeV and above [4]. However, spectrometric parameters not reproduced in all detectors made on the crystals with the same resistivity, carrier mobility and lifetimes. All this range of issues have led us to research, important both from the scientific and applied point of view, aimed at solving a number of physical problems, in particular: (1) investigation of the features of CdTe and $Cd_{1-x}Zn_xTe$ crystals electrical conductivity mechanism, doped with elements of the III or VII groups of the periodic system, taking into account the self-compensation effect; (2) determination of the reasons of low energy resolution of Cd(Zn,Mn)Te-based Ohmic detectors and finding the possibilities to improve the performance of detectors; (3) identify the charge transport mechanisms and the possibilities of reducing the current at high voltages in the X/γ-rays detectors with the Schottky diode in order to improve the device spectrometric characteristics; (4) search for a physical model of quantum efficiency spectral distribution which would explain a significant difference in energy resolution of the ^{55}Fe, ^{241}Am, ^{57}Co, ^{133}Ba, ^{137}Cs isotopes emission spectra taken with CdTe- and $Cd_{1-x}Zn_xTe$-based X/γ-rays detectors; (5) determining the influence of the Schottky diode's space charge region width and the concentration of uncompensated impurities both on the X/γ-rays detectors' quantum efficiency and the charge collection; (6) finding out the possibility of use of CdTe thin polycrystalline films with a barrier structure in detectors with direct X-ray conversion that provides low values of dark current, eliminates the problem of charge collection. Here, we summarize the results of our studies in the above-mentioned areas.

2. Features of conductivity of semi-insulating CdTe and CdZnTe single crystals

This section deals with the results of studies of CdTe crystals, doped with Cl and $Cd_{0,9}Zn_{0,1}Te$ crystals, doped with In, which are widely used for the creation of detectors. The results of the research reveal important features of the electrical characteristics of the crystals. **Figure 1a** presents the typical temperature dependences of the resistivity $\rho(T)$ of the CdTe [5–8] and CdZnTe [9, 10] crystals under study.

The temperature dependence of the intrinsic resistivity of CdTe and CdZnTe crystals $\rho_i(T) = 1/qn_i(\mu_n + \mu_p)$ is also shown for comparison (where де $n_i = (N_cN_v)^{1/2}\exp(-E_g/2kT)$ is the concentration of intrinsic charge carriers, $N_c = 2(m_nkT/2\pi\hbar)^{3/2}$ and $N_v = 2(m_pkT/2\pi\hbar)^{3/2}$ are the effective density of states in the conduction band and the valence band of the semiconductor, respectively). As seen, the values of ρ and ρ_i of both CdTe and CdZnTe are quite close in the whole temperature range, which is important taking into account the necessity to minimize

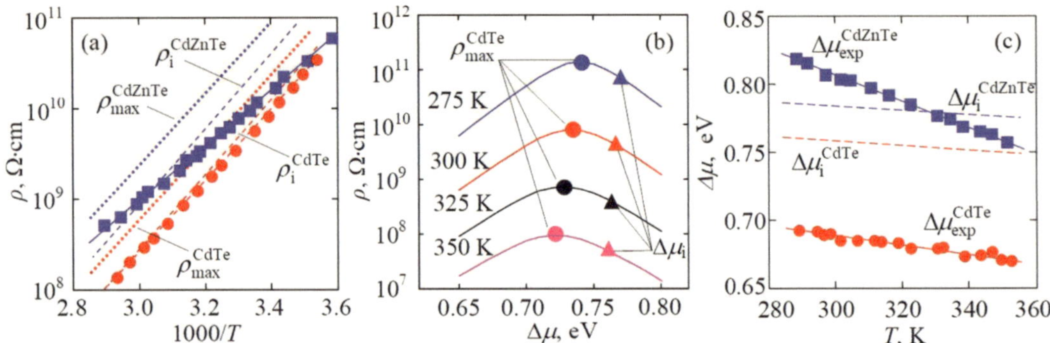

Figure 1. (a) Temperature dependences of resistivity of the CdTe and $Cd_{0,9}Zn_{0,1}Te$ crystals. Dashed lines show the temperature dependences of resistivity of CdTe and $Cd_{0,9}Zn_{0,1}Te$ crystals with intrinsic conductivity ρ_i, dotted lines show its maximum possible values ρ_{max}. (b) Dependence of resistivity of CdTe on position of the Fermi level at different temperatures. The Fermi level energy $\Delta\mu_i$ in intrinsic CdTe is also shown. (c) Temperature dependences of the Fermi level energies in CdTe and CdZnTe crystals. Circles and squares show the values of $\Delta\mu$ (T) calculated with Eq. (1). Solid lines show $\Delta\mu_{calc}(T)$ calculated with Eq. (2). Dashed lines show the Fermi levels in the intrinsic materials.

the dark current in X/γ-rays detectors. At temperatures above the room temperature (>320–330 K), the resistivity ρ of CdTe crystal exceeds its value for a material with intrinsic conductivity ρ_i. At temperatures below ~ 280 K, the resistivity of CdTe crystal exceeds the resistivity of $Cd_{0,9}Zn_{0,1}Te$ with a wider (!) bandgap. The observed excess ρ over ρ_i is explained, the much lower mobility of holes in comparison with the mobility of electrons. If the Fermi level shifts from its position in the intrinsic semiconductor toward the valence band, the contribution of holes into electrical conductivity increases and thus resistivity also increases. However, with further displacement of the Fermi level the resistivity decreases, as the concentration of holes becomes too large. As a result, the thermal activation energy of the electrical conductivity decreases. Giving ρ as $1/q(n\mu_n + n_i^2\mu_p/n)$ and equating to zero the derivative $d\rho/dn$, is easy to show that the maximum value of the resistivity is determined by $\rho_{max} = (2qn_i(\mu_n\mu_p)^{1/2})^{-1}$. As shown in **Figure 1b**, the value of the maximum possible resistivity ρ_{max} significantly exceeds the intrinsic resistivity of CdTe crystal in the whole temperature range. In the case of a semiconductor with almost intrinsic conductivity, the solution of equation $\rho = 1/q(n\mu_n + p\mu_p)$ for Fermi energy $\Delta\mu$ has the form as follows:

$$\Delta\mu = \mathrm{kT}\ln\left(\frac{1 \pm \sqrt{1 - 4q^2\rho^2\mu_n\mu_p n_i^2}}{2q\rho\,\mu_n n_i^2/N_v}\right), \tag{1}$$

where "+" and "–" correspond to n- and p-type semiconductor, respectively. That is, one can find the $\Delta\mu(T)$ dependences of the crystals under study (**Figure 1c**) from the temperature dependence of the resistivity $\rho(T)$ (**Figure 1a**), taking into account the temperature dependences of n_i and mobilities of electrons μ_n and holes μ_p [5]. The Fermi-level energy of the samples calculated with Eq. (1) (**Figure 1c**) shows that the Fermi level is noticeably removed from the conduction band when temperature increase, which slows the growth of the electrons concentration, consequently, leads to a decrease in the of thermal activation energy. As a consequence, despite the fact that the Fermi level located near the middle of the band

http://dx.doi.org/10.5772/intechopen.78504

gap in the whole temperature range, the thermal activation energy is much smaller than that of intrinsic semiconductor. Another important conclusion is that the Fermi level crosses the Fermi level in intrinsic $Cd_{0,9}Zn_{0,1}Te$ at a temperature of ~330 K, that is the material changes the type of conductivity.

An analysis of the statistics of electrons and holes in a semiconductor, in which donor defects form self-compensated complexes, provides additional information on the compensation mechanism in the CdTe and $Cd_{0,9}Zn_{0,1}Te$ crystals. Consideration of the features of self-compensated semiconductors minimizes the number of independent parameters in the calculations [11]. It is important that the concentration of the donor impurity (Cl or In) is significantly higher than the concentration of background impurities and defects. Under this condition, calculations can be made using the simplified scheme of levels in the bandgap, namely, the deep donor and deep acceptor level, as well as the shallow level of donors that did not form complexes, that is, on the "three-level" compensation model [12]. In the electroneutrality equation for semi-insulating wide-band semiconductor with a high concentration of impurities (~10^{18} cm^{-3}), the concentration of free carriers n and p (which do not exceed 10^7 to 10^8 cm^{-3} even at elevated temperatures) can be neglected. It is natural to assume that the level of compensating acceptors is located in the lower part of the bandgap (or at least ~$5kT$ below the Fermi level), that is, they are almost completely ionized, and it can be taken $N_a^+ \approx N_a$, where N_a is the concentration of acceptors. Such acceptors may be Cd or Zn vacancies with ionization energy of 0.43–0.47 eV. With such simplifications, the electroneutrality equation is reduced to the expression $N = N_d^+$, and its analytical solution is

$$\Delta\mu = E_d + \mathrm{kT}\ln\left[\frac{1-\xi}{\xi}\right]. \tag{2}$$

Thus, the Fermi-level energy at a temperature T is determined by the energy of the donor level E_d and its compensation degree $\xi = N_a/N_d$, which can be found by comparing the results of the calculation according to Eq. 2 with the experimental dependences $\Delta\mu(T)$ obtained from the results of measurements of resistivity $\rho(T)$, as shown in **Figure 1c**. The best match of the calculated and experimental dependencies $\Delta\mu(T)$ corresponds to $E_d = 1.081$ eV, $\xi = 0.99998$ for $Cd_{0,9}Zn_{0,1}Te$ and $E_d = 0.783$ eV, $\xi = 0.976$ for CdTe. The obtained values of ionization energy are in the range, in which the photoluminescence bands were detected, which is also consistent with the results of the study of energy levels in the band gap CdTe and $Cd_{1-x}Zn_xTe$ by other methods. Obtained high degree of donor compensation $\xi = 0.979$ for CdTe and $\xi = 0.99998$ for $Cd_{0,9}Zn_{0,1}Te$ confirms the known theoretical fact that an element of Group III or VII of the Periodic system (in this case, Cl and In) as a donor impurity introduced in the crystal lattice, causes the appearance of approximately the same amount of compensating intrinsic defects, which leads to the formation of complexes. It should be emphasized that the obtained values ξ are close to the degree of compensation provided by the Mandel theory for CdTe and ZnTe [13].

These results are important from a practical point of view: (1) At $E_d = 0.60$ eV (as in the crystal $Cd_{0,9}Zn_{0,1}Te$ under study), the values of ρ become close to the maximum value in a very narrow range of ξ (**Figure 2a**). When shifting of the donor to the middle of band gap dome-shaped curve $\rho(\xi)$ expands and its maximum shifts toward values of ξ, close to 0.5. The achievement of the semi-insulating state $Cd_{0,9}Zn_{0,1}Te$ $E_d = 0.60$ eV and $\xi = 0.99998$ became possible due to doping by self-compensated donors and the formation of A- or DX-centers, the concentration of which is practically equal to the concentration of acceptors due to the very nature of these centers. (2) In a certain combination of ionization energy and the compensation degree, changes in the temperature dependence of resistivity and/or inversion of the conductivity type of the crystal in the climate-change temperature range may occur that may affect on the operation of the X/γ-ray detector with the Schottky diode. If the donor level located near the middle of the band gap of the $Cd_{0,9}Zn_{0,1}Te$ crystal (E_d = 0.7 eV), to obtain the resistivity $\rho = 10^{10}$ Ω·cm at 300 K (**Figure 2b**), the compensation degree equals to ξ = 0.679, and the thermal activation energy of conductivity is close to the half of the $Cd_{0,9}Zn_{0,1}Te$ band gap at 0 K (ΔE = 0.84 eV). If E_d = 0.5 eV to ensure $\rho = 10^{10}$ Ω·cm at 300 K, the compensation level should be increased to 0.99979, which leads to a decrease of ΔE to 0.65 eV. When E_d = 0.3 eV, the compensation degree should further increase to 0.9999999, making activation energy decreases to 0.47 eV. If E_d equals to 0.9 eV to ensure $\rho = 10^{10}$ Ω·cm, the compensation degree should be considerably less than 1/2, namely ξ = 0.00047. In this case, the thermal activation energy ΔE will be 1.04 eV, that is, it becomes "abnormally" high [5, 7, 9].

Figure 2c shows the calculated temperature dependences of the Fermi level energy in $Cd_{0,9}Zn_{0,1}Te$ at different compensation degrees of donor (In) with ionization energy E_d = 0.45 eV, which corresponds to the often observed in these crystals photoluminescence band (~1.08 eV). The position of the Fermi level in the intrinsic $Cd_{0,9}Zn_{0,1}Te$ is shown by dashed line. As seen, the Fermi level is located above the Fermi level in the intrinsic semiconductor at the compensation degree ξ = 0.9999 in the whole temperature range, that is, the semiconductor has an n-type of conductivity. At sufficiently higher compensation degree (ξ = 0.999999), the Fermi level is located below the Fermi level in the intrinsic $Cd_{0,9}Zn_{0,1}Te$, that is, the crystal has p-type conductivity [9]. It is also possible the condition when type of conductivity changes in the

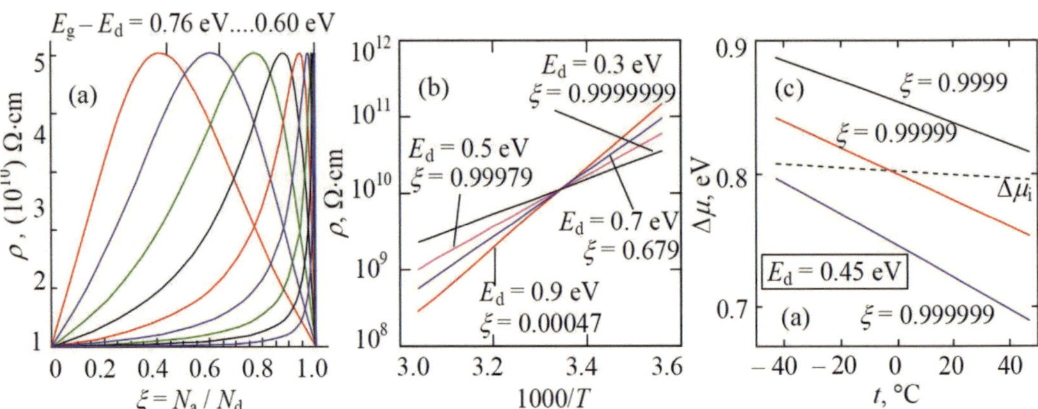

Figure 2. (a) The dependence of resistivity of $Cd_{0.9}Zn_{0.1}Te$ crystal on the compensation degree ξ. (b) Temperature dependences of resistivity ρ at the different ionization energies E_d and compensation degree $\xi = N_a/N_d$, which provides the same ρ, but different thermal activation energies ΔE. (c) The temperature dependences of the Fermi level energy $\Delta\mu$ (T) in $Cd_{0.9}Zn_{0.1}Te$ crystal at the different compensation degrees ($\Delta\mu_i$ - is the Fermi level energy in a crystal with intrinsic conductivity).

climate-change temperature range, which is observed at an intermediate value of $\xi = 0.99999$. Unlike $Cd_{0,9}Zn_{0,1}Te$ with a high compensation degree, in CdTe crystals the compensation degree is much lower and the dependence of the Fermi level on ξ is much weaker. Therefore, in CdTe, the transition from p- to n-type of conductivity occurs at an increase from 0.90 to 0.99, that is, by 9%, while in $Cd_{0,9}Zn_{0,1}Te$—only by 0.01%. This explains how much harder is to grow a homogeneous $Cd_{0,9}Zn_{0,1}Te$ crystal in comparison with CdTe.

3. Charge collection efficiency in CdTe-based Ohmic detectors

The operation of CdTe detector in spectrometric mode assumes complete collection of the charge generated by the absorption of high-energy quanta. Since the lifetime of charge carriers in the most perfect CdTe crystals does not exceed several microseconds, it is necessary to apply a rather high voltage to prevent the "capture" of the carriers by deep impurities (defects). At low voltage applied to the CdTe crystal with ohmic contacts, *I-V* characteristic is linear, but at higher bias a superlinear increase in current is always observed [14]. Attention is drawn to the fact that the deviation from linear *I-V* relationships at higher voltage is observed stronger when the temperature decreases (**Figure 3b**, inset). Therefore, we can assume that an additional charge transport mechanism with much weaker temperature dependence comes into play with increasing voltage. This is confirmed by the data in **Figure 3b**, which show the voltage dependences of the difference ΔI between the measured current I and a linearly extrapolated current I_o ($\Delta I = I - I_o$). As seen, the current excessive over linearly extrapolated current is virtually independent of temperature. A deviation of *I-V* characteristics from linearity due to lowering the barrier at imperfect Ohmic contact should be rejected because such a mechanism leads to an exponentially increase in the current with temperature [15]. The current caused by tunneling transitions of electrons from the Fermi level in the metal (or slightly below it) into the semiconductor can be almost temperature independent [16]. However, at bias voltage in the range 10–100 V, the probability of tunneling is practically zero. A much greater probability of tunneling is through a thin interfacial oxide layer, whose presence on the crystal surface before metal deposition cannot be excluded (**Figure 3c**). At higher voltages, the linear behavior of the *I-V* characteristic of the CdTe crystal is replaced by a quadratic dependence on V, as in the case of space charge limited current (SCLC)] according to the Mott-Gurney law [17]. It is confirmed by comparing the experimental data with the extrapolation of the quadratic *I-V* dependence (**Figure 3b**). According to theory, SCLC can be formed by injection of charge carriers into the valence or conduction band. In the case of semi-insulating CdTe, the injection of electrons into the conduction band should be preferred. Firstly, the excess concentration of electrons above the electron equilibrium concentration is achieved much easier, since in the CdTe:Cl crystals the electron concentration approximately two orders of magnitude lower than that of holes. Second, the electron mobility in CdTe is more than an order of magnitude higher than that of holes, and the SCLC is proportional to the charge carrier mobility. Finally, the electron current injected by tunneling is almost temperature independent, that is, observed from the experience and just this one needs to explain. Taking into account the current of thermally generated holes and SCLC we can write:

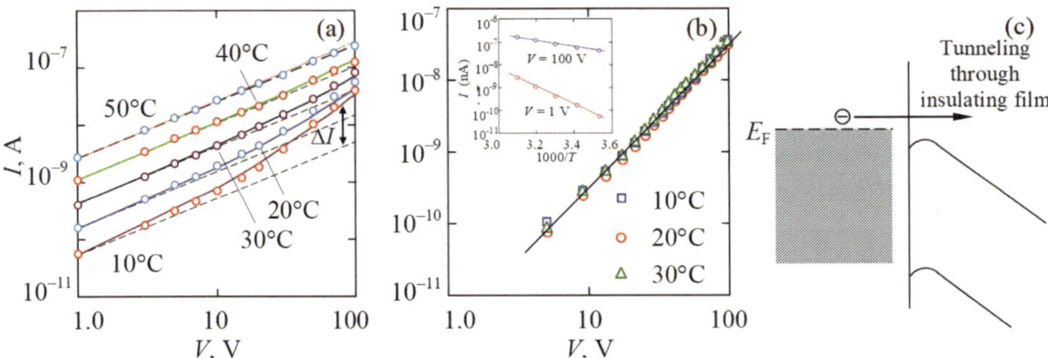

Figure 3. (a) I-V characteristics of CdTe crystal measured (circles) and calculated using Eq. (3) at different temperatures (solid lines). The dashed lines show linear extrapolation of I-V dependencies at low V. (b) Voltage dependence of difference ΔI between the measured current I and a linearly extrapolated current I_o for three temperatures. The straight solid line extrapolates quadratic dependence of the current on voltage. The inset shows temperature dependences of the currents at 1 and 100 V. (c) Energy diagram of the metal/CdTe contact showing tunneling of electrons through an intermediate oxide film on the CdTe crystal surface.

$$I = sqp_o(T)\,\mu_p \frac{V}{d} + sK\frac{9}{8}\frac{\varepsilon\varepsilon_o \mu_n}{d^3}V^2, \quad (3)$$

where ε is the dielectric permittivity of the semiconductor, ε_o is the dielectric constant of vacuum, μ_n and μ_p is the mobility of charge carriers, s is the area of the contact, p_0 is the equilibrium concentration of holes in the crystal, K is the coefficient, which takes into account the probability of electron tunneling through the oxide film and reducing the contribution of the SCLC.

Temperature variation leads to a change in current through the crystal, depending on the SCLC contribution. For the best agreement, the calculation results by Eq. (3) with the experimental data we should substitute $K = 2.3{\cdot}10^{-4}$. As seen, Eq. (3) well reproduces the voltage dependences of the current and its temperature changes in detail (**Figure 3a**). SCLC negatively impacts on the detector performance since it leads to an increase in leakage current, and hence degrades the energy resolution of the detector. Moreover, the contribution of the SCLC increases with decreasing temperature due to increasing the holes mobility (by 5–6% per 10°C) (**Figure 4a**). Such a character of the SCLC does not allow reduce significantly the leakage current by thermoelectrically cooling of the detector as in the case of CdTe detectors with Schottky diode [18].

SCLC is proportional to the squared voltage and is inversely proportional to the crystal thickness d; therefore, the relationship between the CdTe crystal thickness d and the drift length of carriers λ_p increases with decreasing d (**Figure 4a**, inset), which, in turn, improves the energy resolution of detector. Thus, too low energy resolution of X/γ-rays CdTe-detector with two Ohmic contacts (6–8%) caused by ineffective charge collection. One of the reasons might be the recombination of charge carriers in the bulk and on the front and back surfaces of the crystal. Analysis of the influence of surface recombination can be made on the base of continuity equation (with the corresponding boundary conditions) and taking into account the drift and diffusion components of the current. The calculations show that for the actual thicknesses

http://dx.doi.org/10.5772/intechopen.78504

of the crystal and the applied voltages, the recombination on the surfaces of the crystal can be neglected. Neglecting recombination losses at the crystal surfaces, we have to assume that the low efficiency of charge collection in Ohmic-type CdTe detectors is caused by trapping of photogenerated charge carriers by deep levels of impurities (defects) in the crystal bulk. These losses is strongly dependent on the lifetime of charge carriers (τ_n and τ_p), which at a given electric field F (together with their mobility (μ_n and μ_p) determine the drift length of carriers $\lambda_n = \mu_n F \tau_n$, $\lambda_p = \mu_p F \tau_p$. In this view, the relationship between the CdTe crystal thickness and the drift length of carriers (**Figure 4a**, inset) is very important. A quantitative description of the collection losses of photogenerated charge carriers gives the well-known Hecht equation, which for a uniform electric field has the form [19].

$$\eta_H(x) = \frac{\lambda_n}{d}\left[1 - \exp\left(-\frac{d-x}{\lambda_n}\right)\right] + \frac{\lambda_p}{d}\left[1 - \exp\left(-\frac{x}{\lambda_p}\right)\right]. \tag{4}$$

Taking into account the most important processes which determine the spectral distribution of the quantum detection efficiency, in particular, absorption in the bulk of crystal and an electrode material (Pt), the losses caused by trapping of charge carriers in the bulk of crystal, the detection efficiency depending on the absorption coefficient α_γ in the crystal with Ohmic contacts can be written as

$$\eta(\alpha_\gamma) = \int_0^d T_{\mathrm{Pt}}(\alpha_\gamma)\, \alpha_\gamma \exp(-\alpha_\gamma x)\, \eta_H(x) \mathrm{d}x, \tag{5}$$

where $T_{\mathrm{Pt}}(\alpha_\gamma)$ takes into account the radiation attenuation after passing through an electrode material; $\eta_H(x)$ is Hecht function (4); $\alpha_\gamma \exp(-\alpha_\gamma x)$ is the generation rate of electron–hole pairs per incident photon [16].

Insufficient absorptive capacity in high spectral range significantly reduces the registration of X/γ-rays but does not affect the processes that take place after photon absorption. Therefore,

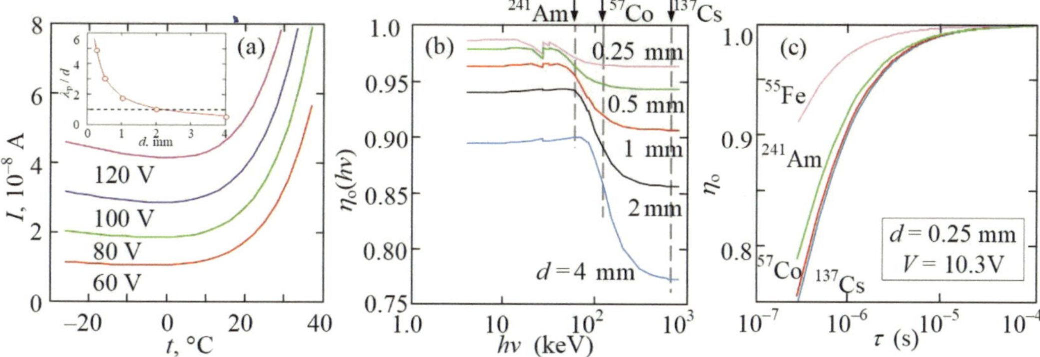

Figure 4. (a) Temperature dependences of the current at different voltages applied to the CdTe crystal. The inset shows the dependences of the drift length of holes on the thickness of the CdTe crystal at voltage when the same leakage current of 3 × 10^{-8} A is achieved. (b) Charge collection efficiency spectra for CdTe crystals of different thicknesses at voltages that corresponds to the same current 3 × 10^{-8} A. (c) Energy resolution in the spectra of different isotopes for the detector thickness of 0.25 mm at voltage of 10.3 V.

the spectral distribution of the charge collection efficiency (the value determining the energy resolution of the detector) is obtained by dividing the detection efficiency $\eta(h\nu)$ by the absorption capacity of the crystal $A(h\nu) = 1 - \exp(-\alpha_\gamma d)$. **Figure 4b** shows the collection efficiency curves $\eta_0(h\nu)$ calculated for the different thicknesses of CdTe and voltages at which the current is the same as at voltage of 60 V for crystal thickness of 1 mm ($3{\cdot}10^{-8}$ A). As seen, when the crystal thickness is 4 mm, the charge collection efficiency $\eta_0(h\nu)$ in the photon energy $h\nu < 100$ keV is 90%, while for $h\nu \approx 1$ MeV $\eta_0(h\nu)$ is reduced to 77% (**Figure 4b**). When the crystal thins, the charge collection efficiency is significantly improved reaching a level of 97–98%. With the thickness of 0.25 mm the charge collection efficiency is above 97% throughout the whole spectral range. However, increasing the charge collection efficiency $\eta_0(h\nu)$ with thinning of the crystal is achieved with a significant decrease in the efficiency of detection (registration) in the range of high-energy photons. A significant increase in the energy resolution can be achieved by improving the quality of CdTe crystals and, thus, increasing the lifetime of charge carriers. Our calculations show that with increasing the electron lifetime by the order of magnitude from 3×10^{-6} s to 3×10^{-5} s, for the crystal thickness of 0.25 mm at voltage that corresponds to the current 3×10^{-8} A (10.3 V) the energy resolution in the spectra of all isotopes is higher than 99% [20] (**Figure 4c**).

CdZnTe and CdMnTe-based X/γ-rays Ohmic detectors can have electrical characteristics both similar to presented above and different. This is largely due to the choice of contacts material, the treatment of the crystal surface before contacts fabrication, conditions of post-deposition treatment. At low voltage applied to the p-$Cd_{1-x}Mn_xTe$ (x = 0.3) crystal *I-V* characteristics are linear, but at higher bias a superlinear increase in current is observed approximately the same extent at different temperatures. The fact that the voltage dependence of difference between the measured current and a linearly extrapolated current is quadratic, which indicates that the observed supernular growth of current is due to by space charge limited current (SCLC) according to the Mott-Gurney law [17]. The activation energy of the conductivity caused by the equilibrium holes (at V = 10 V) equals to 0.39 eV. Attention is drawn to the fact that the energy of acceptor trap at the formation of the SCLC in the Ni/CdMnTe contact (at V > 200 V), equal also 0.39 eV, that is, impurity (or defect), responsible for electrical conductivity of material and trap of injected charge carriers clearly have the same nature. Therefore, the same activation energy for the current of equilibrium holes and the current surplus of equilibrium current confirms the fact that SCLC in the Ni/CdMnTe/Ni detector is formed by the injection of majority carriers (holes) from the metal, not by the tunnel injection of minority carriers (electrons) as in the case of Pt/CdTe/Pt detectors discussed above.

$Cd_{1-x}Zn_xTe$ (x = 0.1) n-type crystals with gold Ohmic contacts show other features of superlinear current growth at high voltages. At voltages, lower ~ 10 V *I-V* characteristic are linear, but at higher bias, the superlinear increase is observed. However, the voltage of deviation from the linear law is 10–20 V regardless of temperature. It turns out that the current of equilibrium electrons and the excess current in the Au/CdZnTe/Au detector are growing approximately equally with the temperature. This is confirmed by the fact that the thermal activation energy of the crystal $Cd_{0,9}Zn_{0,1}Te$ is 0.74 eV, and the thermal activation energy of the excess current at

http://dx.doi.org/10.5772/intechopen.78504

100 V is also quite high $\Delta E = 0.65$ eV, which causes its significant growth from temperature. The voltage dependence of the current, surplus of equilibrium current, found by extrapolation of the linear part of the *I-V* characteristic at low voltages has a complex form. In the voltage range $V = 20–40$ V the current is rapidly increasing, at high voltages (up to the highest voltages), the power $\Delta I \sim V^{2,4–2,5}$ is observed. This behavior of the Au/CdZnTe/Au detector characteristics contradicts the SCLC theory. Therefore, the reason for the deviation of the current from the linear dependence is the injection of minority charge carriers (holes) due to the imperfection of Ohmic contacts.

4. Electrical characteristics of Schottky diodes based on semi-insulating CdTe single crystals

The section deals with electrical characteristics of Ni/CdTe/Ni X/γ-rays detectors with Schottky diodes based on high-resistivity CdTe single crystals ($\rho \sim 10^9$ Ω·cm (300 K)).

The theoretical analysis of experimental results allows identifying and explaining the essential features of the charge transport mechanisms depending on the resistivity of the material and the parameters of the diode structure, in particular the concentration of uncompensated impurities (defects) and the height of the potential barrier on Schottky contact [21]. According to the Sah-Noyce-Shockley theory, the current through the diode is determined by the integration of the generation-recombination rate over the whole space charge region (SCR) width [22].

$$I_{g-r} = Aq \int_0^W \frac{n(x,V)p(x,V) - n_i^2}{\tau_{po}[n(x,V) + n_1] + \tau_{no}[p(x,V) + p_1]} dx, \tag{6}$$

where A is the diode area, q electron charge, W is the width of the SCR, $n(x,V)$ and $p(x,V)$ - are the concentrations of charge carriers in the conduction and valence bands, respectively, τ_{no} and τ_{po} - are the effective lifetimes of electrons and holes in the SCR, and the quantities $n_1 = N_c\exp(-E_t/kT)$ and $p_1 = N_v\exp[-(E_g - E_t)/kT]$ are determined by the depth of the generation-recombination level E_t. The results of calculations of the *I–V* characteristic, by using formula (6) show that the model of generation-recombination processes in the SCR adequately describes not only the current dependence on the voltage, but also the temperature induced variations in the Ni/p-CdTe Schottky diode *I–V* characteristic: (1) The reverse current, which has a generation origin, cannot vary in a wide range of the material resistivity ρ since this current is governed by the carrier lifetime and by the thickness of the SCR, which have no direct relation with a value of ρ. (2) In the region of low forward biases, where the dependence $I \propto \exp(qV/2kT) - 1$ holds, the current is governed by the same parameters and, therefore, is also only slightly ρ-dependent. (3) As ρ increases, the Fermi level recedes from the valence band; that is, $\Delta\mu$ increases at the same time as φ_0 decreases. In this case, the part of the forward branch, where the forward current is proportional to $\exp(qV/2kT)$, is increasingly restricted from above, as is observed in the experimental curves.

The Ni/CdTe/Ni diode structure with Schottky and near-Ohmic contacts at the CdTe(111)A and CdTe(111)B surfaces of semi-insulating CdTe single crystals (ρ = (2–4)·10^9 Ω·cm) demonstrates absence of rectification properties at bias voltages lower than 6–7 V, which can be attributed to a very high resistance of the CdTe substrate, that is, the voltage drop across the bulk part of the crystal should be taken into account. It should be noted that consideration of the voltage drop has strongly modified the shape of the forward *I-V* characteristic of the studied diode structure (**Figure 5b**) [21, 23]. A sharp increase in the current at higher forward bias voltages is attributed to the injection of minority carriers from the Schottky contact into the neutral part of the crystal and the modulation of its electrical conductivity, which is confirmed by the results of calculations. It should be emphasized that the Ni/CdTe/Ni detectors with Schottky and near-Ohmic contacts demonstrates low reverse current (~10^{-9} A/cm^2 at 300 K) at high reverse bias due to significant bending on the Ni/CdTe Schottky contact and low enough level of minority carrier injection from the near-Ohmic CdTe/Ni contact into the neutral part of the diode structure. It should be noted, the generation-recombination Sah-Noyce-Shockley theory analytically describes the *J-V* characteristic of the diode structure at different temperatures (**Figure 5b**) [24, 27]. Analysis of the voltage dependence of the differential resistance R_{diff} shows, at forward connection decreases with increasing in the low-bias region (**Figure 5c**). In the voltage range V = 1–3 V, the R_{diff} saturates, which means that the energy barrier is practically compensated by applied voltage and further voltage drop takes place across the bulk part of the diode structure. With further increasing the forward bias voltage, a sharp decrease in R_{diff} is observed. The value of R_{diff} becomes 2–3 orders of magnitude less than resistance of the bulk part of the diode R_s. Such lowering of R_{diff} is explained by injection of electrons (minority carriers) from the forward-biased Schottky contact into the bulk part of the crystal and modulation of its resistance (**Figure 5a** and **c**). Indeed, at higher forward voltage the barrier φ_o lowers and electron injection in the bulk part of the crystal is increasingly enhanced.

Analysis of the reverse *J-V* characteristic at high-bias voltages that is most important and interesting in the application of CdTe diodes as X/γ-ray detectors (V < 600–700 V) shows that the reverse current through the diode structure is controlled by the reverse-biased Schottky contact. A sublinear rise in the current (it is typical for the generation charge transport mechanism) corresponds to a gradual increase in the differential resistance (**Figure 5c**). However, on exceeding 600–700 V, the differential resistance decreases increasingly and then steeply decays at similarly to that at forward connection of the diode at voltages higher than a few volts. It can be explained by injection of electrons from the near-Ohmic contact into the bulk of the crystal [21, 23, 24] (**Figure 5a**). With an increase in the current, a fraction of the applied voltage, much like for a forward connection of the device, drops across the neutral part of the crystal and only a small its fraction drops across the near-Ohmic contact on the opposite side of the crystal. Thus, we have come to not at all trivial conclusion that at relatively high reverse bias, the processes in the "Ohmic" contact affect the reverse-biased Schottky contact on the opposite side of the crystal [15, 21, 23, 24]. A decrease in injection of carriers from the near-Ohmic contact in a Schottky diode with Ni/CdTe/Ni electrode configuration is an important way to reduce the leakage current and improve the performance of CdTe based X/γ-ray detectors. On increasing the operating voltage at low-leakage current allows to enhance the detection efficiency of the device especially in the region of high energy of photons.

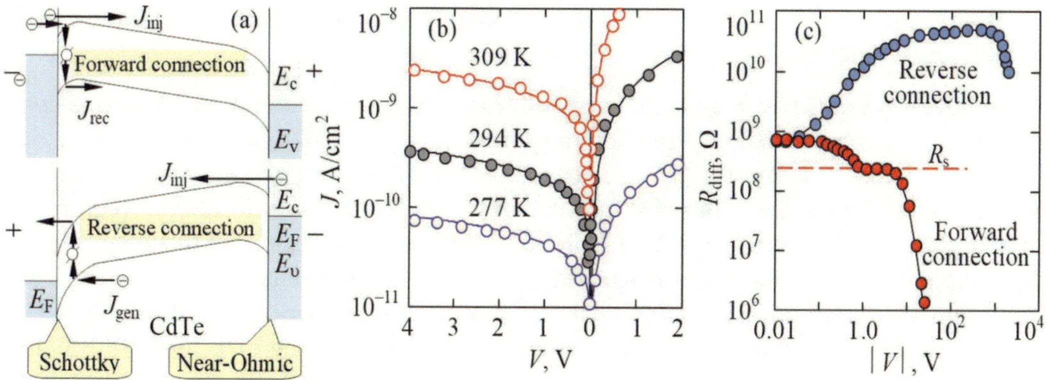

Figure 5. (a) The energy diagram of the forward and reverse biased Ni/CdTe/Ni diode structure shown at the top and bottom, respectively. The recombination (J_{rec}), generation (J_{gen}) an injection (J_{inj}) currents are shown by arrows. (b) *J-V* characteristics of the Ni/CdTe/Ni structure at different temperatures. The circles show the measurement results; the lines are the results of calculations by Eq. (6). (c) Differential resistance of the detector in a wide range of forward and reverse biased. The dashed straight line shows the resistance of the bulk part of the diode structure R_s.

5. Detection efficiency of CdTe based X/γ-ray detector

The parameters of crystal and diode structure significantly affect the quantum detection efficiency and energy resolution of detectors based on semi-insulating CdTe and $Cd_{0,9}Zn_{0,1}Te$ crystals with Schottky diode. In such crystals with deep levels of impurities (defects) in the band gap, the density of the space charge and the intensity of the electric field grow rapidly near the crystal surface, enhanced with the increase in the degree of compensation of the semiconductor (in contrast to the Schottky diodes on the semiconductor with shallow impurities levels). Minority charge carriers play an insignificant role in the formation of space charge, despite the presence of an inverse layer near the surface of the semiconductor. In spite of the features of the formation of SCR in the Schottky diodes based on self-compensating semiconductors (which are CdTe and $Cd_{0,9}Zn_{0,1}Te$ crystals, doped with Cl or In), the difference between the value of the SCR width, as determined by the solution of the Poisson equation, and by means of the known the formula for the Schottky diode does not exceed 15–16% even with a high compensation degree, that is, the width of the SCR is quite accurately determined by the concentration of uncompensated impurities. The charge collection efficiency in X/γ-ray detectors with a Schottky diode essentially depends on the carrier lifetime τ. It is important for practice that the charge collection efficiency is noticeably lower than 1 when the lifetime is less than 10^{-8} s, whereas to provide practically the total charge collection (99%) in the Ohmic detector the carrier lifetime should be equal to or exceed ~10^{-6} s.

The resistivity of CdTe and CdZnTe crystals under study at room temperature are (2–3)·10^9 and (3–5)·10^{10} Ω·cm, respectively. The band gap of the crystals E_g for CdTe equals to 1.47–1.48 eV, for $Cd_{0,9}Zn_{0,1}Te$ $E_g = 1.53$ eV at room temperatures. Our studies of the relaxation curves of the rise and decay of the photocurrent excited by rectangular pulses of semiconductor laser ($\lambda = 782$ nm) showed that the lifetimes of the charge carriers in the SCR and in the neutral part of the CdTe crystal differ significantly. In the case of CdTe crystal with two Ohmic contacts, the lifetimes of electrons amount to a few microseconds, which is consistent with the data presented on the site of Acrorad Co. Ltd. [14]. If the crystal is irradiated through a semitransparent Schottky contact, the laser radiation is absorbed in a thin near-surface layer of the SCR and lifetimes of carriers are relatively short (10–20 ns).

Although all crystals had high resistivity and minority carrier lifetime, the diodes showed significant differences in the registration of spectra from ^{137}Cs (662 кеВ), ^{133}Ba (356 кеВ), ^{57}Co (122 кеВ), ^{241}Am (59 кеВ), ^{55}Fe (5,9 кеВ) isotopes. The CdTe detector was a high resolution detector, however the CdZnTe registered the spectra but with lower resolution (**Figure 6a**). At first glance it seems unclear as CdTe inferior in characteristics $Cd_{0,9}Zn_{0,1}Te$. Obviously, the detecting properties of the diode structure are influenced by other characteristics of the material. There is an assumption that such a parameter is the SCR width of a Schottky diode, which in a compensated semiconductor can be significant. Indeed, in the detector with Schottky diode, the SCR itself is an active area of the detector, and its width, of course, is one of the most important parameters. Therefore, in compensated semiconductor in addition to the high resistivity and long carrier lifetimes that is necessary for high detection efficiency, another mandatory requirement to the concentration of uncompensated impurities in the material (which determines the SCR width of a Schottky diode) is substantiated.

The detection efficiency spectra of CdTe-based crystals with Schottky diode taking into account the drift and diffusion components can be expressed as [25].

$$\eta = \frac{\lambda_n}{W}\left[1 - \exp\left(-\frac{W}{\lambda_n}\right)\right]\left(\int_0^W \alpha_{\mathrm{CdTe}} \exp(-\alpha_{\mathrm{CdTe}}\, x)\mathrm{d}x + \frac{\alpha L_n}{1+\alpha L_n}\exp(-\alpha W)\right). \tag{7}$$

Figure 6b shows the detection efficiency spectra of CdTe ($Cd_{0,9}Zn_{0,1}Te$) crystals with Schottky diode at the voltage $V = 400$ V applied to the detector and different concentrations of uncompensated donors $N_d - N_a$ in the material. The results clearly illustrate the fact that the spectra of $\eta(h\nu)$ can significantly be modified when $N_d - N_a$ is changed. If $N_d - N_a$ decreases from 10^{14} to 10^{10} cm^{-3}, the detection efficiency of ^{55}Fe ($h\nu = 5.9$ keV) and ^{241}Am ($h\nu = 59.5$ keV) isotopes vary almost by 3 and 2 orders of magnitude, respectively. At the same decreasing N_d - N_a, the detection efficiency of ^{57}Co ($h\nu = 122$ keV) isotope varies within one order of magnitude and the detection efficiency of ^{133}Ba (356 keV) and ^{137}Cs (662 keV) isotopes vary relatively weak. An important feature of the results is that the dependences $\eta(N_d–N_a)$ for all the isotopes are described by a curve with maximum (**Figure 6c**) [6, 10, 11, 25]. As seen, in all cases, the detection efficiency rather rapidly increases as the SCR widens starting at high uncompensated impurity concentrations (10^{15} cm^{-3}). In addition, recombination losses in the SCR also increase and ultimately become so significant that the detection efficiency decreases with a further increase in $N_d - N_a$. The obtained results for the measurements and calculations show that, together with high resistivity, lifetime and mobility of charge carriers, the concentration of uncompensated impurities in the range 10^{11}–10^{13} cm^{-3} can be considered also necessary condition for the efficient operation of X/γ-rays detectors based on CdTe and $Cd_{0,9}Zn_{0,1}Te$ [10, 11]. It is the concentration of uncompensated impurities of 10^{12} cm^{-3} in CdTe crystals made it possible to obtain ^{137}Cs radioisotope energy spectrum by an Ni/CdTe/Ni diode detector at applied reverse bias voltage of 1200 V with the record values of energy resolution at room temperature (2.8 keV of FWHM at 662 keV) (**Figure 6c**, inset).

http://dx.doi.org/10.5772/intechopen.78504

To determine the concentration of uncompensated impurities in crystals of CdTe and $Cd_{0,9}Zn_{0,1}Te$, we compared the detection efficiency with irradiation of the crystal by the Ohmic contact side and the Schottky contact side. In the high-energy region of the spectrum, in which the absorption coefficient for X/γ-rays (α_γ) is small, excitation occurs virtually uniformly over the entire crystal volume and the detection efficiency for a crystal with the Schottky contact is independent of which side of the detector is irradiated, the side of the Schottky contact or the side of the Ohmic contact. This is confirmed experimentally. If a CdTe detector with a Schottky contact is subjected to the radiation of the ^{137}Cs isotope, the peak height at the photon energy 662 keV ($\alpha_\gamma \approx 0.1$ cm^{-1}) is practically the same when different sides of the sample are irradiated, the side of the Schottky contact or the side of the Ohmic contact. If the isotope ^{55}Fe is used ($\alpha_\gamma \approx 4000$ cm^{-1}), the peak height in the case of irradiation of the Schottky contact side is by two orders of magnitude, than that in the case of irradiation of the Ohmic contact side, as is shown in **Figure 6a** (inset) [11]. This is accounted for by the fact that, at $\alpha_\gamma \approx 4000$ cm^{-1}, the effective depth of radiation penetration into the material is smaller than 1 μm, therefore, in the case of irradiation of the Ohmic contact side, a significant portion of electrons, which appeared as a result of γ-photon absorption, do not reach the SCR by diffusion [17]. Evidently, in this case, the peak height greatly depends on the SCR width and, consequently, on the concentration of uncompensated impurities in the semiconductor, which can be used to determine the value of $N_d - N_a$. Thus, there is a significant difference between the concentrations of uncompensated impurities in CdTe and $Cd_{0,9}Zn_{0,1}Te$ crystals, which are used in the fabrication of X/γ-rays detectors. The concentration of uncompensated impurities is ~(1–3) × 10^{12} cm^{-3} for CdTe crystals and (1–5) × 10^{8} cm^{-3} (i.e., four orders of magnitude lower) for $Cd_{0,9}Zn_{0,1}Te$ crystals [11]. The low concentration of uncompensated impurities (10^8–10^9 cm^{-3}) is the reason for the unsatisfactory detectivity of $Cd_{0,9}Zn_{0,1}Te$ detectors regardless of a fully acceptable resistivity of crystals (> 10^9 Ω·cm) and the lifetime of charge carriers (> 10^{-6} s) [10].

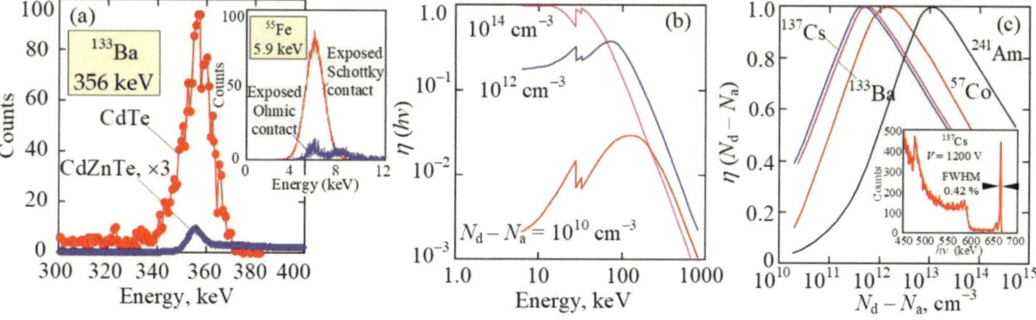

Figure 6. (a) Spectra of ^{133}Ba isotope taken with Schottky diode detectors based on CdTe and at V = 500 V. The inset shows the emission spectra of the ^{55}Fe isotope measured by a CdTe detector with a Schottky contact under irradiation of different sides of the sample. (b) Detection efficiency spectra of CdTe-based detector with Schottky diode calculated for different concentrations of uncompensated donors $N_d - N_a$. (c) Normalized detection efficiency of different isotopes as a function of $N_d - N_a$. The inset shows the typical ^{137}Cs radioisotope energy spectrum detected by an Ni/CdTe/Ni diode detector.

6. CdTe-based Schottky diode X-ray detectors for medical imaging

In the section the possibilities of using diode structures based on relatively low resistivity p-CdTe and n-CdTe ($\rho = 10^3$–10^4 Ω·cm), and polycrystalline CdTe in direct-conversion digital flat-panel X-ray image detectors are discussed.

Investigation of electrical properties, charge collection processes and the spectral distribution of the detection efficiency of X/γ-ray detectors based on CdTe with relatively low resistivity and Schottky contact confirm that their characteristics are generally inferior to technologically more complicated detectors with diodes Schottky based on semi-insulating CdTe. In particular, the currents of Schottky diodes under study are rather small for this type of diode structures—about 1 nA with a contact area of 3.5 mm^2. The current are determined by the generation-recombination in the SCR according to the Sah-Noyce-Shockley theory [22, 26].

The total detection efficiency for a detector utilizing a Schottky diode is the sum of the drift and diffusion components [27, 28]. As shown in **Figure 7a**, the contribution of the diffusion component to the total efficiency of the detector is quite important at $\tau_p = 10^{-6}$ s and in the case of high-energy photons (i.e., at lower absorption coefficients) it is dominant. The efficiency of charge collection of the X/γ-rays detector with the Schottky diode substantially depends on the lifetime of charge carriers τ and concentration of uncompensated donors N_d - N_a (**Figure 7b**, inset). If N_d - N_a value is ~10^{14} cm^{-3}, in order to ensure a practically complete charge collection (>99%), the lifetime of the carriers should equal or exceed ~10^{-7} s. If N_d - $N_a \approx 10^{16}$ cм$^{-3}$, the lifetime of charge carriers should not exceed 10^{-9} s, which is quite real when using even poor quality CdTe crystals. For X-ray examination of a breast, in the photon energy region $h\nu < 30$ keV n-type CdTe detector is more acceptable (**Figure 7b**). However, for X-ray examination of a chest ($h\nu > 50$ keV) p-CdTe should be chosen (**Figure 7c**). The detection efficiency of X-ray in the Al/p-CdTe diode structure at the maximum possible electron lifetime (a few microseconds) is 50–70 and 20–40% in the photon energy ranges of 20–30 and 50–80 keV, respectively (**Figure 7c**). Such characteristics seem to be acceptable for mammography and chest radiography.

The use of a stacked CdTe detector with a Schottky diode, which is already practiced, can significantly improve the detecting efficiency of the device especially in the high-energy range of the spectrum. In the energy range ~100 keV the efficiency of a stacked detector is greater than that of a single layer detector [28]. The highly developed technology of the deposition of polycrystalline CdTe layers of large area with a surface-barrier structure in solar cells can be adapted to the fabrication of flat-panel X-ray image detectors. The presence of a barrier structure in the relatively low resistivity CdTe ($\rho = 10^4$–10^6 Ω·cm) provides a low-leakage (dark) currents comparable with those in a-Se photoconductors ($\rho = 10^{12}$ Ω·cm at 300 K). In a CdTe diode structure, virtually full charge collection occurs independently of the applied voltage at the carrier lifetime $\tau > 10^{-7}$ s and uncompensated impurity concentration higher than 10^{14} cm^{-3}. Electric field concentration in the space charge region of a barrier structure eliminates the problem of the collection of charge generated by X-ray photon absorption (there is no need to increase the operating voltage up to several kiloelectronvolts as in the case of a-Se photoconductors).

http://dx.doi.org/10.5772/intechopen.78504

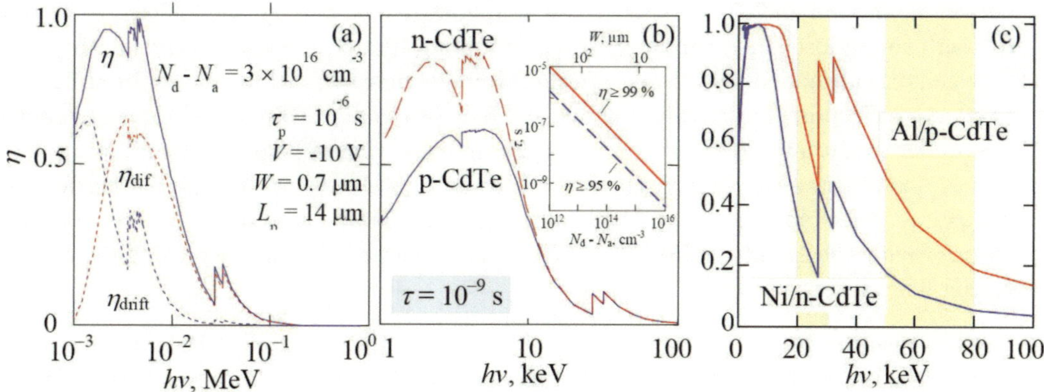

Figure 7. (a) The drift η_{drift} and diffusion η_{dif} components of the detection efficiency in the Schottky diode and their sum η. (b) Comparison of the total detection efficiency of the n- and p-CdTe-based Schottky diodes with the carrier lifetime 10^{-9} s. The inset shows the correlation between the charge-carrier lifetime τ and concentration of uncompensated donors $N_d - N_a$ at which the collection of 99 and 95% charge carriers photogenerated at the interface between the depleted region and neutral part of the diode structure ($x = W$) is achieved at $V = -100$ V. (c) The detection efficiency of the Ni/n-CdTe and Al/p-CdTe diodes calculated for the uncompensated impurity concentration of 10^{16} cm^{-3} and carrier lifetime $\tau = 3 \times 10^{-6}$ s. The photon energy ranges which are used for diagnostics of breast and chest are shown by shading.

7. Conclusions

1. The unconventional peculiarities, which are important from scientific and practical points of view, have been revealed by the experimental studies of the temperature dependences of resistivity and Fermi level energy of semi-intrinsic $Cd_{0,9}Zn_{0,1}Te$:In and CdTe:Cl crystals used for fabrication of X/γ-ray detectors: (i) the material can be semi-insulating when the compensation degree of a deep impurity level located near the middle of the band gap is around 0.5 (in this case, the Fermi level is pinned); (ii) if the impurity level is not close to the middle of the band gap, the semi-insulating condition is reached at low or high compensation degree (in this case, the Fermi level position strongly depends on the temperature T and activation energy ΔE can be significantly higher than one-half of the band gap at $T \rightarrow 0$ K as it takes place in an intrinsic semiconductor). Among other things, this can lead to inversion of the conductivity type of the semiconductor as the temperature varies during climatic operation of a device that leads to qualitative changes of the electric properties of both Schottky and Ohmic contacts in X/γ-ray detectors.

2. The features of operation of X/γ-ray detectors based on Cd(Zn, Mn)Te crystals with two Ohmic contacts have been established and ways to improve their energy resolution have been determined by comprehensive investigation of the electrical characteristics and quantum efficiency of detectors: (i) in the CdTe detector a rapid rise of the current with increasing voltage higher than 6–8 V for the crystal thickness of 1 mm is caused by the SCLC. A distinctive feature of the current is its temperature independence because the mechanism of injection of charge carriers is tunneling through the thin insulating film between the crystal and metal contact; (ii) thermoelectric cooling, commonly used for CdTe Schottky diode detectors, does not provide the desired result since it leads to a decrease in leakage

current no more than 2–3 times when the temperature lowers from 300 to 260–270 K, and further cooling loses its meaning; (iii) with thinning the semiconductor crystal, the ratio between the carrier drift length and crystal thickness increases that improves the efficiency of charge collection at relatively low bias voltage; (iv) in the CdZnTe and CdMnTe detectors under study, a rapid rise of the current with increasing voltage and temperature due to injection of charge carriers is observed, so these crystals are not suitable for fabrication of X/γ-ray detectors; (v) a significant increase in the energy resolution can be achieved by improving the quality of Cd(Zn, Mn)Te crystals and, as a result, increasing the charge carrier lifetime.

3. The key results have been obtained from the investigation of the electrical characteristics of the Ni/CdTe/Ni structures with a Schottky contact based on CdTe:Cl crystals with nearly intrinsic conductivity: (i) the *I-V* characteristics of the Schottky diode structure with low reverse leakage current at high-bias voltages can be quantitatively described in terms of the known physical models: the generation-recombination in the SCR, the processes under conditions of strong electric fields and currents limited by space charge; (ii) a rapid rise of the current at high direct voltages due to injection of minority carriers from the Schottky contact to the neutral part of the crystal and the modulation of its conductivity; (iii) at relatively high reverse bias, the processes in the "Ohmic" contact on the opposite side of the crystal affect the reverse-biased Schottky contact.

4. The investigation results, obtained for Schottky diode detectors based on CdTe and $Cd_{0.9}Zn_{0.1}Te$ crystals with high resistivity (~10^9–10^{10} Ω·cm) and minority carrier lifetime (~10^{-6} s) and demonstrating significant differences in detection of the spectra from ^{241}Am, ^{57}Co, ^{133}Ba and ^{137}Cs isotopes, have shown a correlation between the low concentration of uncompensated donors and poor detection efficiency of the $Cd_{0.9}Zn_{0.1}Te$ detectors with quite acceptable resistivity and carrier lifetime. The conducted measurements and calculations show that the concentration of uncompensated impurities in the range from 3×10^{10} to 3×10^{12} cm^{-3} is yet another obligatory condition for effective operation of X/γ-ray Schottky diode detectors based on CdTe and $Cd_{1-x}Zn_xTe$ crystals.

5. The important conclusions about application of the CdTe layers with a Schottky contact in direct-conversion flat-panel X-ray image detectors have been established based on the research results of the electrical and detection characteristics of the fabricated diode structures: (i) the highly developed technology of the deposition of polycrystalline CdTe layers of large area with a surface-barrier structure in solar cells can be adapted to the fabrication of flat-panel X-ray image detectors. The presence of a barrier structure in the relatively low resistivity CdTe ($\rho = 10^4$–10^6 Ω·cm) provides low-leakage (dark) currents comparable with those in a-Se photoconductors ($\rho = 10^{12}$ Ω·cm 300 K); (ii) electric field strength in the space charge region of a barrier structure eliminates the problem of the collection of charge generated by X-ray photon absorption. In a CdTe diode structure, virtually full charge collection occurs independently of the applied voltage at the carrier lifetime $\tau > 10^{-7}$ s and uncompensated impurity concentration higher than 10^{14} cm^{-3}; (iii) the detection efficiency of X-rays in the Al/p-CdTe diode structure at the maximum possible electron lifetime (a few microseconds) is 50–70 and 20–40% in the photon energy ranges

of 20–30 and 50–80 keV, respectively. Such characteristics seem to be acceptable for mammography and chest radiography.

Acknowledgements

This research was supported by the Collaborative Project COCAE (Grant SEC-218000) of the European Community's Seventh Framework Programme and by the Collaborative Project SENERA (Grant SfP-984705) of the NATO Science for Peace and Security Programme. The authors express gratitude to all colleagues indicated as co-authors in Refs. [5-11, 15, 20, 21, 23-28] for their contribution in carrying out the investigations.

Author details

Olena Maslyanchuk[1]*, Stepan Melnychuk[1], Volodymyr Gnatyuk[2] and Toru Aoki[2]

*Address all correspondence to: emaslyanchuk@yahoo.com

1 Yuriy Fedkovych Chernivtsi National University, Chernivtsi, Ukraine

2 Research Institute of Electronics, Shizuoka University, Hamamatsu, Japan

References

[1] Szeles C. CdZnTe and CdTe materials for X-ray and gamma ray radiation detector applications. Physica Status Solidi B. 2004;**241**(3):783-790. DOI: 10.1002/pssb.200304296

[2] Sordo SD, Abbene L, Caroli E, Mancini AM, Zappettini A, Ubertini P. Progress in the development of CdTe and CdZnTe semiconductor radiation detectors for astrophysical and medical applications. Sensors. 2009;**9**:3491-3526. DOI: 10.3390/s90503491

[3] Triboulet R, Siffert P. CdTe and Related Compounds; Physics, Defects, Hetero-and Nano-Structures, Crystal Growth, Surfaces and Applications. Amsterdam, The Netherlands: Elsevier; 2009, 550 p. ISBN: 9780080914589

[4] Takahashi T, Mitani T, Kobayashi Y, Kouda M, Sato G, Watanabe S, Nakazawa K, Okada Y, Funaki M, Ohno R, Mori K. High-resolution Schottky CdTe diode detector. IEEE Transactions on Nuclear Science. 2002;**49**:1297-1303. DOI: 10.1109/TNS.2002.1039655

[5] Kosyachenko LA, Diequez E, Maslyanchuk OL, Melnychuk SV, Sklyarchuk OV, Sklyarchuk OF, Grushko EV, Aoki T, Lambropoulos CP. Special features of conductivity of semi-intrinsic CdTe and CdZnTe single crystals used in X- and γ-ray detectors. Proceedings of SPIE. 2010;**7805**:78051-78059. DOI: 10.1117/12.862726

[6] Kosyachenko L, Aoki T, Lambropoulos C, Gnatyuk V, Sklyarchuk V, Maslyanchuk O, Grushko E, Sklyarchuk O, Koike A. High energy resolution CdTe Schottky diode γ-ray detectors. IEEE Transactions on Nuclear Science. 2013;**60**(4):2845-2852. DOI: 10.1109/TNS.2013.2260356

[7] Kosyachenko LA, Aoki T, Maslyanchuk OL, Melnychyk SV, Sklyarchuk VM, Sklyarchuk OV. Features of conduction mechanism of semi-insulating CdTe single crystals. Semiconductors. 2010;**44**(6):699-704. DOI: 10.1134/S1063782610060023

[8] Maslyanchuk OL, Kosyachenko LA, Melnychuk SV, Fochuk PM, Aoki T. Self-compensation limited conductivity of Cl-doped CdTe crystals. Physica Status Solidi C. 2014;**11**(9):1519-1522. DOI: 10.1002/pssc.201300694

[9] Kosyachenko LA, Melnychuk SV, Maslyanchuk OL, Sklyarchuk VM, Sklyarchuk OF, Fiederle M, Lambropoulos CP. Self-compensation limited conductivity in semi-insulating indium-doped $Cd_{0,9}Zn_{0,1}Te$ crystals. Journal of Applied Physics. 2012;**112**: 013705(1-7). DOI: 10.1063/1.4733463

[10] Kosyachenko LA, Lambropoulos CP, Aoki T, Diegues E, Fiederle M, Loukas D, Sklyarchuk OV, Maslyanchuk OL, Grushko EV, Sklyarchuk VM, Crosso J, Bensalah H. Concentration of uncompensated impurities as a key parameter of CdTe and CdZnTe crystals for Schottky diode x/γ-ray detectors. Semiconductor Science and Technology. 2012;**27**:015007 (1-11). DOI: 10.1088/0268-1242/27/1/015007

[11] Kosyachenko LA, Sklyarchuk VM, Melnychuk SV, Maslyanchuk OL, Grushko EV, Sklyarchuk OV. Effect of uncompensated impurity concentration on properties of CdTe-based X- and γ-ray detectors. Semiconductors. 2012;**46**(3):374-381. DOI: 10.1134/S1063782612030153

[12] Fiederle M, Eiche C, Salk M, Schwarz R, Benz KW, Stadler W, Hofmann DM, Meyer BK. Modified compensation model of CdTe. Journal of Applied Physics. 1998;**84**(12):6689-6692. DOI: 10.1063/1.368874

[13] Mandel G. Self-compensation limited conductivity in binary semiconductors. I. Theory. Physical Review. 1964;**134A**:1073-1079. DOI: 10.1103/PhysRev.134.A1073

[14] Acrorad Co, Ltd., 13-23 Suzaki, Gushikawa, Okinawa 904-2234, Japan. http://www.acrorad.co.jp/us

[15] Kosyachenko LA, Gnatyuk VA, Aoki T, Sklyarchuk VM, Sklyarchuk OF, Maslyanchuk OL. Super high voltage Schottky diode with low leakage current for X- and γ-ray detector application. Applied Physics Letters. 2009;**94**:092109-092111. DOI: 10.1063/1.3093839

[16] Sze SM, Physics of Semiconductor Devices, 2nd ed. Wiley-Interscience, New York, 1981. ISBN: 0-471-09837-X

[17] Lampert MA, Mark P. Current Injection in Solids. New York and London: Academic Press; 1970. 363 p. ISBN: 0124353509

[18] Amptek Inc., 14 De Angelo Drive, Bedford, MA, 01730, USA; http://www.amptek.com/xr100cdt.html

[19] Hecht K. Zum Mechanismus des lichtelektrischen Primrstromes in isolieren den Kristallen. Zeitschrift für Physik. 1932;**77**(3-4):235-245. DOI: 10.1007/BF01338917

[20] Aoki T, Maslyanchuk OL, Kosyachenko LA, Gnatyuk VA. Reasons of low charge collection efficiency in CdTe-based X/γ-ray detectors with ohmic contacts. Proceedings of SPIE. 2013;**8852**:1-13. DOI: 10.1117/12.2022657

[21] Kosyachenko LA, Sklyarchuk VM, Sklyarchuk OF, Maslyanchuk OL, Gnatyuk VA, Aoki T. Higher voltage Ni-CdTe Schottky diodes with low leakage current. IEEE Transactions on Nuclear Science. 2009;**56**(4):1827-1835. DOI: 10.1109/TNS.2009.2021162

[22] Sah C, Noyse R, Shokley W. Carrier generation and recombination in p-n-junction and p-n-junctions characteristics. Proceedings of the IRE. 1957;**45**(9):1228-1243. DOI: 10.1109/JRPROC.1957.278528

[23] Kosyachenko LA, Maslyanchuk OL, Sklyarchuk VM. Special features of charge transport in Schottky diodes based on semi-insulating CdTe. Semiconductors. 2005;**39**(6):722-729. DOI: 10.1134/1.1944866

[24] Kosyachenko LA, Maslyanchuk OL, Sklyarchuk VM, Grushko EV, Gnatyuk VA, Aoki T, Hatanaka Y. Electrical characteristics of semi-insulating CdTe single crystals and CdTe Schottky diodes. Journal of Applied Physics. 2007;**101**(1):013704(1-9)). DOI: 10.1063/1.2401283

[25] Kosyachenko LA, Aoki T, Lambropoulos CP, Sklyarchuk VM, Grushko EV, Maslyanchuk OL, Sklyarchuk OV. Optimal width of barrier region in X/γ-ray Schottky diode detectors based on CdTe and CdZnTe crystals. Journal of Applied Physics. 2013;**113**:054504(1-9). DOI: 10.1063/1.4790358

[26] Gnatyuk VA, Vlasenko OI, Kosyachenko LA, Maslyanchuk OL, Sklyarchuk VM, Grushko EV, Aoki T. CdTe-based Schottky diode X-ray detectors for medical imaging. Proceedings of SPIE. 2008;**7008**:1-10. DOI: 10.1117/12.797360

[27] Kosyachenko LA, Maslyanchuk OL, Gnatyuk VA, Lambropoulos C, Rarenko IM, Sklyarchuk VM, Sklyarchuk OF, Zakharuk ZI. Charge collection properties of a CdTe Schottky diode for X- and γ-rays detectors. Semiconductor Science and Technology. 2008;**23**:075024-075031. DOI: 10.1088/0268-1242/23/7/075024

[28] Kosyachenko LA, Maslyanchuk OL. Efficiency spectrum of a CdTe X- and γ-ray detector with a Schottky diode. Physica Status Solidi C. 2005;**2**(3):1194-1199. DOI: 10.1002/pssc.200460661

Chapter 4

Understanding Low-Dose Exposure and Field Effects to Resolve the Field-Laboratory Paradox: Multifaceted Biological Effects from the Fukushima Nuclear Accident

Joji M. Otaki

Additional information is available at the end of the chapter

http://dx.doi.org/10.5772/intechopen.79870

"Everything should be made as simple as possible, but not simpler."

--- *Albert Einstein*

Abstract

Many reports about the biological effects of the Fukushima nuclear accident on various wild organisms have accumulated in recent years. Results from field-based laboratory experiments using the pale grass blue butterfly have clearly demonstrated that this butterfly is highly sensitive to "low-dose" internal exposure from field-contaminated host-plant leaves. These experimental results are fully consistent with the filed-collection results reporting high abnormality rates. In contrast, this butterfly is highly resistant against the internal exposure to chemically pure radioactive cesium chloride under laboratory conditions. To resolve this field-laboratory paradox, I propose that the field effects, which are a collection of indirect effects that work through different modes of action than do the conventional direct effects, play an important role in the "low-dose" exposure results in the field. In other words, exclusively focusing on the effects of direct radiation, as predicted by dosimetric analysis, may be too simplistic. In this chapter, I provide a working definition and discuss the possible variation in the field effects. I include an example on the misunderstanding of the field effects In the United Nations Scientific Committee on the Effects of Atomic Radiation (UNSCEAR) 2017 Report. Lastly, I discuss a theoretical application of the butterfly model to humans.

Keywords: biological effect, ecological effect, field effect, Fukushima nuclear accident, low-dose exposure, indirect effect, UNSCEAR

1. Introduction

In terms of economic loss, the Fukushima nuclear accident that occurred in 2011 and the Chernobyl nuclear accident that occurred in 1986 are the worst nuclear accidents in the history of mankind [1]. Although considerable research results have accumulated for the Chernobyl disaster, there are still considerable debates concerning its biological effects [2–4]. The reasons for these disagreements among researchers are likely multifaceted, but one reason stems from the fact that the Chernobyl nuclear accident occurred in the former Soviet Union, which made it difficult for international researchers to easily access the contaminated areas and the critical data. In addition, some important tools and methods for biological analyses, such as those for genomic analysis and computational applications, were not yet available at that time. Considering these points, the Fukushima nuclear accident is the first historical case in which researchers have been politically and technically allowed to perform field work and laboratory experiments after such a major nuclear accident. In other words, scientists working in the second decade of the twenty-first century are responsible for correctly evaluating the biological effects of the Fukushima nuclear accident.

Because of the large-scale nature of the accident, many research questions have been developed for studies on the biological consequences of the accident at the ecological, organismal, and molecular levels [5]. However, the most important question is to determine *how severe the biological impacts from the accident are*. This is different from questions that investigate how severe the biological impacts from radiation exposure (or, more precisely, effective radiation doses) are. That is, the direct impacts from the exposure to radiation are possibly only one type of the impacts from the accident. However, many researchers have tried to understand the biological impacts of the Fukushima nuclear accident by exclusively studying the effective doses based on radiation dosimetry. And dosimetric data are often exclusively used for risk assessment and management. The idea behind this approach is that radioactivity (and its direct exposure) is the sole "pollutant" that causes any biological impacts. There is no question that radiation doses are important; however, this cannot justify the exclusion of other factors that may cause more powerful effects on biological systems.

Another important presumption of using the dosimetric approach to determine biological impacts is that researchers completely understand the system in question (at least at first), enabling a precise level of prediction of the biological impacts that often reference the recommendations and mathematical simulations of the International Commission of Radiological Protection (ICRP) (e.g., [6–8]). That is, it is presumed that reference levels of the effects of radiation exposure on certain organisms such as humans are completely known and these references are credible and applicable to the case of the Fukushima nuclear accident. It is to be understood that the reference levels are just for protection purposes only as experience-based values to balance risk and benefit for residents, patients, workers, and researchers. Nonetheless, it can be said that dosimetric predictions mostly take a *we-know-all approach* regardless of researchers' awareness. Although there are many studies that support these reference levels, some dosimetric studies for the Fukushima nuclear accident often lack efforts to perform or incorporate field and laboratory studies that look for possible phenotypic and genetic effects; in other words, these studies often appear to conclude that such field and laboratory experiments are not necessary because the system has already been known well.

In contrast, the biological and ecological approach may be called a *we-know-little approach*; in other words, the biological impacts of the accident reflect the things that we do not know well, and these are the topics that should be evaluated in field work and laboratory experiments in the real world. For example, studies using the biological approach may admit that organisms face many different stress conditions in the wild, and they are found in unique positions in the ecological network; as a result, sometimes unexpected consequences in terms of an organism's response to pollutants may be observed.

As discussed in the next section, we have been using *the pale grass blue butterfly* for Fukushima research since 2011; research began immediately after the Fukushima nuclear accident [9]. One of the most important types of experiments in the research on the pale grass blue butterfly in Fukushima is the so-called the internal exposure experiment. In this experiment, the field-collected host-plant leaves, which are contaminated at various levels (judged by the radiation levels of ^{137}Cs and ^{134}Cs), were given to butterfly larvae collected from the least-contaminated area, i.e., Okinawa (approximately 1700 km southwest of the Fukushima Dai-ichi Nuclear Power Plant); these experiments resulted in high mortality and abnormality rates [9–12]. These results support the field reports of high rates of abnormality [9, 13–15]. A mutagenesis study of this butterfly produced similar phenotypes [16], and the effects on body size detected in the first paper [9] were also supported by the field and experimental results [17]. Although these field and experimental results may be surprising in light of the conventional view of radiation biology and physics, the experimental procedures were rigorous enough to support these conclusions [18].

Furthermore, in recent years, many field reports have accumulated on the possible effects on various organisms [5], and these are consistent with our results. Such studies include the bird and arthropod populations [19–21], gall-forming aphids [22], Japanese monkey [23, 24], barn swallow [25], goshawk [26], rice plant [27, 28], fir tree [29], red pine tree [30], and intertidal species populations including the rock shell [31]. Furthermore, the possible changes induced by the nuclear accident have been reported at the biochemical level. For example, stress responses in cattle may have been induced in contaminated areas [32]. Changes in gene expression have been reported in the small intestine of pigs [33]. Other reported cases include DNA damage in bovine lymphocytes [34], enhanced spermatogenesis [35], and chromosomal aberrations [36, 37] in large Japanese field mice; however, there are reports in which mammalian testes collected from bull, bore, Inobuta, and large Japanese field mice in the contaminated area did not show any noticeable abnormalities [38–40].

In contrast, one of the most recent results of ours came from a series of similar internal exposure experiments in which radioactive ^{137}Cs was supplied to larvae as a form of chemically pure cesium chloride solution in an artificial diet; however, the results have not yet been published. It is likely that the pale grass blue butterfly is highly resistant to internal irradiation alone, as expected from the conventional understanding of insects' high resistance to irradiation. This discrepancy between the two systems may be called *the field-laboratory paradox*. The difference between the two systems is clear. The former system used contaminated leaves from the real world, while the latter system used an "ideal" pure source of cesium chloride in an artificial diet. I conclude that the latter system is not entirely relevant to the case of the Fukushima nuclear accident, and the former system may be heavily influenced by several different modes of the indirect field effects that are not well known to researchers. A similar situation has already appeared in mammals and aphids. An experiment on internal ^{137}Cs irradiation in mice did not indicate any detectable change in the litter size and sex ratio [41]; in contrast, at least some of the field data have suggested adverse effects in mammals, as discussed above. Striking morphological abnormalities of aphids reported from the polluted areas [22] were not reproduced in the process of embryogenesis and egg hatching by irradiation experiments, although a change in developmental time was detected [42].

Precise dosimetric analysis of larvae may provide additional information that satisfies dosimetrists; however, dosimetric analysis does not play a major role in reaching the conclusions stated above if the radioactivity concentration of the diet is known to us. What is important is the fact that the same experimental system was employed in studies of the pale grass blue butterfly; the two experiments simply used different types of food, i.e., either the field-harvested contaminated leaves or the artificial diet containing ^{137}Cs. Moreover, the results from the former experiment are fully supported by the field work. Based on the butterfly case and the mammalian case discussed above, this kind of field-laboratory paradox is likely widespread among organisms of various taxa. Indeed, a literature survey showed that the controlled laboratory effects and field effects were very different in terms of their sensitivity levels; the field cases from Chernobyl were eight times more sensitive than the laboratory-controlled external irradiation cases [43].

Undoubtedly, dosimetric analysis provides a different level of insight. For example, the inferred genetic mutations that are heritable over generations in this butterfly [9] are likely caused by the high-level acute exposure immediately incurred after the accident rather than by the low-level chronic exposure [12, 44, 45]. To evaluate these effects, it is important to dosimetrically understand the absorbed doses of the butterfly at the initial time of the event.

In this chapter, I will discuss several important issues associated with "low-dose" radiation exposure and field effects; additionally, I propose the importance of non-dosimetric studies in conjunction with conventional dosimetric studies. Borrowing the famous phrase from Shakespeare's *Hamlet*, researchers who engage in the biological consequences of the Fukushima nuclear accident may consider the following: "*To be or not to be* (i.e., dosimetry), *that is the question.*" However, the answer is clear: both approaches are necessary to advance this scientific field to a higher level. In other words, the final answer to this question is "*to be and not to be.*" I believe that this is the only way to reveal a holistic picture of the biological impacts of the Fukushima nuclear accident, which would serve as a basis for risk assessment and management of nuclear pollution.

2. The pale grass blue butterfly: a versatile indicator

Multiple biological approaches should be used to understand the real-world phenomena resulting from the Fukushima nuclear accident. Furthermore, to understand biological phenomena in general, it is customary for biologists to concentrate on a few *surrogate species* or *model species*. For example, in developmental genetics, the fruit fly *Drosophila melanogaster* is an important model species [46]. In conservation biology, many types of surrogate species are often proposed, including indicator, umbrella, keystone, and flagship species, to evaluate the quality of the natural environment [47]. The simultaneous use of multiple indicator species from different taxonomic groups is generally favorable [47] but may be difficult in practice. To understand biological impacts of the Fukushima nuclear accident, studies that use *indicator species* are likely required.

If only a single (or a few) species is used in biological studies of the Fukushima nuclear accident, the pale grass blue butterfly is one of the ideal systems of choice in that it is associated with (and almost dependent on) the living environment of humans; as a result, the butterfly reflects the health of the human environment [12, 44, 45, 48]. Using this butterfly, efficient field work can be performed, and relatively fast and precise experiments can be performed in the laboratory [49, 50]. Other advantages of using this butterfly have been discussed elsewhere [12, 44, 45, 48].

It should be noted that using nonhuman model organisms to obtain information relevant to humans is not a novel approach in biomedical sciences. In fact, it is a common practice to use the fruit fly and even yeast to infer the molecular mechanisms of human diseases. The fruit fly is used not because it is the invertebrate most similar to humans but because it is practically useful for experimental manipulation. This model organism approach to human-related research is valid because, at the molecular level, there are many commonalities among organisms. Furthermore, as discussed in Taira et al. [12] and Otaki [48], radiation effects are molecular events. DNA may be damaged "directly" by radiation or "indirectly" by other ionized molecules, such as water (note that the usage of "direct" and "indirect" here is different from the terminology discussed in most parts of this chapter). The molecular-level ionizing mechanisms are universal in all organisms, including humans and this butterfly species. In this sense, the butterfly data are applicable to humans. Likely, the unconventional field effects that are discussed below may also occur in universal molecular events. Thus, the field effects that were detected in the butterfly are also likely applicable to humans, at least to some extent; however, the precise mechanistic understanding of the field effects on the molecular events is still unclear.

In contrast to the uniform molecular markings found in many organisms, the manifestation of these effects (i.e., phenotypic effects) may be very different among species. In butterflies, morphological abnormalities such as leg and wing deformation are relatively frequent; however, no suitable counterpart of this phenotypic effect can be identified in humans. Such organismal-level phenotypic effects (i.e., disease manifestations) in humans are not readily inferable from butterfly data.

3. Targeted and nontargeted effects

The dosimetric approach often states that ionizing radiation targets DNA directly or indirectly through the ionization of water molecules (hence, they are called targeted effects) and that the degree of DNA damage is linearly reflected in the biological consequences. These statements mean that biological effects can be predicted by the effective dose. Although this approach is widely accepted and utilized for assessing the biological impacts of nuclear disasters, the approach entirely ignores other potential molecular pathways and dismisses the complexity of the biological and ecological responses to the various known and unknown materials that are released from nuclear reactors.

In contrast to the conventional targeted effects, the last two decades have experienced a surge of *nontargeted effects* of ionizing radiation [51–56]. The nontargeted effects include bystander effects, genomic instability, adaptive responses, and other modes, and these nontargeted effects are likely caused by the reactive oxygen species produced by irradiation [51–56]. In this sense, the nontarget effects may be referred to as the indirect effects (note that in this chapter, nontargeted effects are classified into the same category as the direct effects as a matter of convenience to some extent). In terms of the nontargeted effects, it is important to remember that they are not readily predictable by doses, and many of them are latent. Therefore, the nontarget effects may not be detected in acute irradiation experiments, but they may manifest in the field. Furthermore, the field-laboratory paradox discussed above may have originated, at least partly, from the influence of the nontargeted effects in the field. In fact, the nontargeted effects, such as genomic instability, may have played significant roles in the observed increase in butterfly morphological abnormalities in the fall of 2012 [9, 13].

However, even the nontarget effects may not adequately explain the all effects that manifest in the field. For example, there could be possible nonradioactive by-products released from a reactor and naturally occurring nonradioactive materials that are "activated" by the radioactive materials released from a reactor. There may also be ecological interactions that could amplify small irradiation effects to larger levels throughout a food web. These possibilities may be potential sources of the *field effects* (or more precisely, *field-specific effects*), which would not be observed in controlled laboratory experiments that use an artificial source of radiation, such as ^{60}Co and chemically pure ^{137}Cs. However, these field-specific effects should not be confused with (or dismissed as) confounding factors because these field effects are elicited by the nuclear accident. Similarly, nontargeted effects do not have to be field-specific effects; nontargeted effects may be observed in controlled laboratory experiments that use an artificial radiation source and a simple biological system, such as a cell culture system. In other words, the nontargeted effects may be uncovered with conventional radiation biology, which investigates universal mechanisms of radiation effects, but the field-specific effects may be uncovered with *pollution biology*, which investigates the real-world phenomena; however, these two fields cannot be separated in a meaningful way in the case of nuclear accidents.

In this chapter, I refer to both the conventional targeted effects and the nontargeted effects as the "direct" effects (or "primary" effects) (**Figure 1**); however, in some literature, the nontargeted effects or one mode of the nontargeted effect are referred to as the indirect effects. It is understood that laboratory-based controlled irradiation experiments, irrespective of high or low doses, primarily examine the direct effects of ionizing radiation. In contrast, as mentioned above, other potential unconventional indirect effects of nuclear pollution are collectively called the field effects (**Figure 1**) [48, 57]. The field effects are often dependent on a biological (including ecological) context.

http://dx.doi.org/10.5772/intechopen.79870

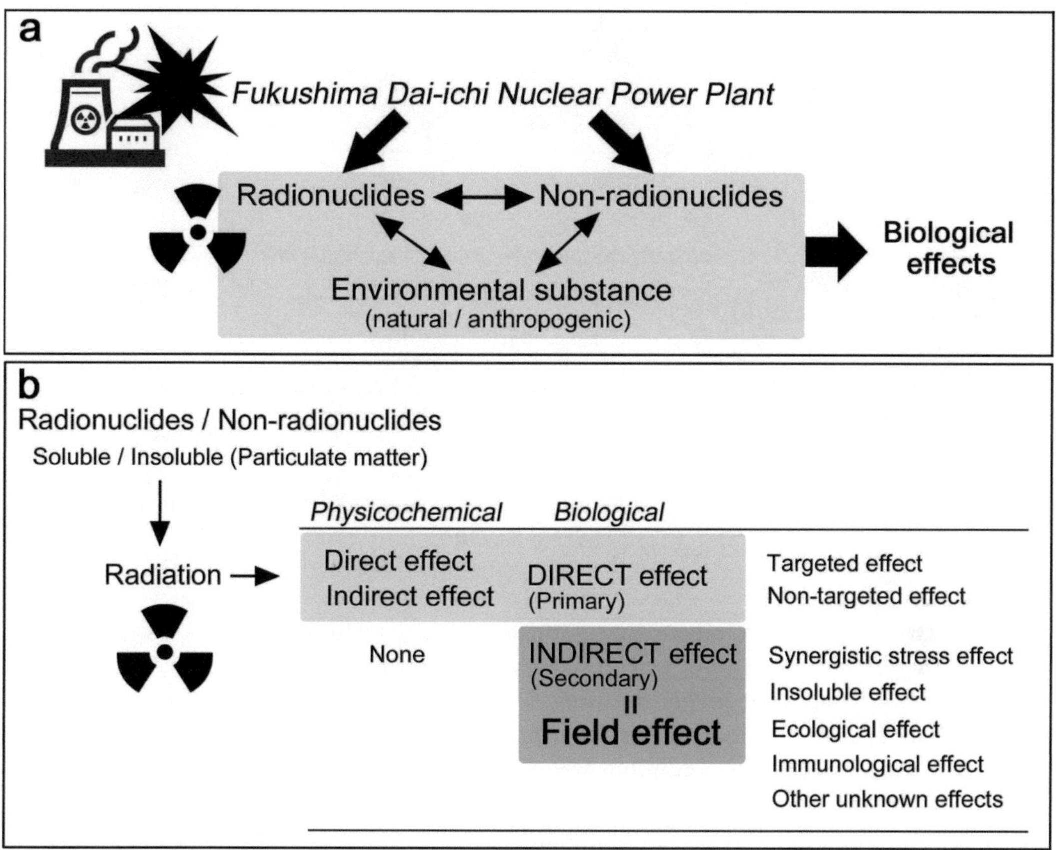

Figure 1. Possible effects of the explosion of the Fukushima Dai-ichi Nuclear Power Plant. (a) Overall pathways. The Fukushima Dai-ichi Nuclear Power Plant released radionuclides as well as non-radionuclides that may not be fully identified. They interact with each other, and they also interact with environmental substances. Environmental substances could be natural (biotic or abiotic) or anthropogenic. The collective outputs of these interactions manifest as biological effects. The illustration of a nuclear power plant was obtained from a free illustration site called Icon-rainbow (http://icon-rainbow.com/). (b) Multifaceted radiation effects. Released substances may be radionuclides or non-radionuclides, and they may be soluble or insoluble as particulate matter. Physicochemically, ionizing radiation has direct or indirect effects on the major biological target, i.e., DNA. However, both types of effects may be considered as biological direct (or primary) effects. In contrast, there are multiple biological indirect (i.e., secondary) effects, depending on the context from which the organism in question faces. The latter is often field-specific, and thus called the field-specific effects (or simply the field effects).

4. Field effects (1): synergistic effects

The biological indirect effects are a collective expression of all biological effects of the nuclear accident excluding the effects of the direct radiation exposure. Because any wild biological system has diverse and complex relationships with biological and chemical species, there are numerous indirect pathways that can affect organisms. Below, the field effects are roughly categorized into three groups: synergistic effects, effects from particulate matters, and ecological effects (**Figure 2**).

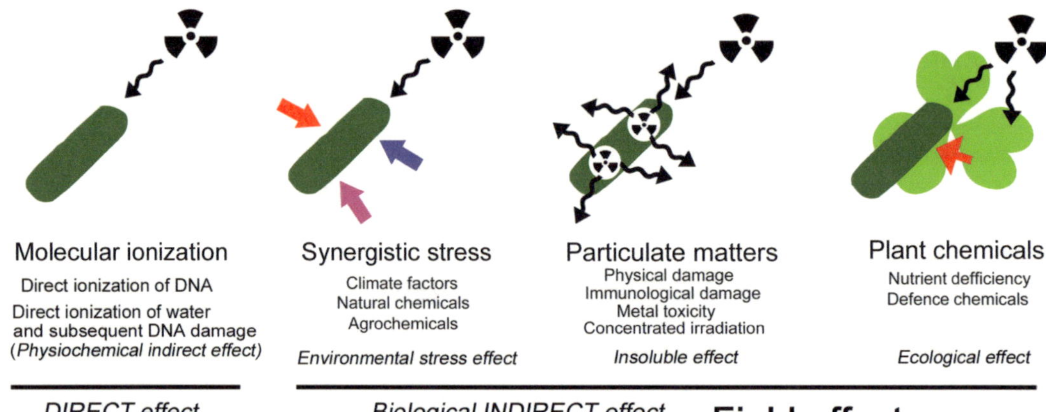

Figure 2. Four possible types of effects on the larvae of the pale grass blue butterfly (green bars). Molecular ionization is the direct (i.e., primary) effect, while the other three modes (synergistic stress, particulate matter, and plant chemicals) are biological indirect (i.e., secondary) field effects.

First, *synergistic effects* with other environmental factors, including climate conditions and chemical stressors, may exist in the wild. When an organism experiences stress from a single source, the stress may be managed relatively well; however, when stress is imposed by two different sources, the harmful effects may be synergistically enhanced beyond their individual actions. In laboratory conditions, the "climate" conditions are usually constant, and additional stressors are not usually provided; thus, synergy is often difficult to predict using conventional irradiation experiments alone. Logically, the synergistic effects of radiation exposure and other stressors have been an important topic in radiation biology [52, 53, 58–61]. However, in my opinion, such synergistic stress effects have not been fully appreciated in radiation biology. Importantly, synergistic stress effects are not limited to exposure to radiation. Here, I briefly discuss two examples that may be insightful for this line of discussion.

A discrepancy has been recognized between the laboratory and field results in phenotypic plasticity studies. In an authoritative textbook, Gilbert and Epel [62] stated the following: "Phenotypic plasticity means that animals in the wild may develop differently than those in the laboratory" and "This has important consequences when we apply knowledge gained in the laboratory to a field science such as conservation biology." One specific example provided in the textbook states that some frog tadpoles are up to 46 times more sensitive to pesticides in the presence of predators that release chemicals in the wild than they are in the laboratory [63, 64]. The conclusion stated that "ignoring the relevant ecology can cause incorrect estimates of a pesticide's lethality in nature" [63]. I believe that the same principle applies to radioactive materials from nuclear reactors.

Another insightful case was reported in the epidemic caused by the bacterium *Clostridium difficile* [65, 66]. For this bacterial epidemic outbreak to occur in North America and Europe, the widespread use of a food additive, trehalose, played a crucial role. Infected mice had higher mortality rates when fed food that contained trehalose [66]. Without the trehalose-rich environment that newly emerged in this century, the deadly endemic would not have occurred.

Although trehalose alone may not be a significant stressor, this case illustrates an example of an unexpected synergistic interaction between toxic substances that were otherwise benign environmental chemicals.

5. Field effects (2): particulate matter

Second, what was released from the Fukushima nuclear reactors was a plume of materials that caused *particulate air pollution;* regardless of whether these particulates were radioactive, the released materials were dispersed as atmospheric aerosols [67, 68]. There is no question that atmospheric aerosols cause respiratory and cardiovascular diseases in humans [69–72]. Indeed, natural radon attaches to air dust, and when this dust is inhaled, it is believed to cause lung cancer [73]. There is no reason to believe that the particulate air pollution from the nuclear reactors was safe for butterflies or other wild organisms. However, to my knowledge, any discussion from this viewpoint is scarce.

It should be noted that the plume from the nuclear reactors contained two types of radioactive materials: soluble and insoluble forms. Soluble materials, such as a form of inorganic salt, are solubilized quickly in environmental water. Additionally, insoluble materials have been detected as spherical particles [74, 75], and they are attached on the surface of any material. At least some of these particles (i.e., particulate matter) may bind to nonradioactive common air dust [68, 69]. Based on the results of the internal exposure experiments in which field-collected polluted leaves were fed to butterfly larvae, the ingestion of particulate matter present on the surface of leaves may have caused digestive and immunological effects [9–12].

6. Field effects (3): ecological effects

Third, when one examines the interactions of multiple species based on a food web or an ecological system as a whole, one may be able to discover radiation effects that would not be discovered by a single-species approach; consequently, observations like this may indicate important field effects. This may be called *the ecological effects*. A similar concept has recently been addressed in radioecology [76]; however, this topic is often discussed from the viewpoint of the bioaccumulation of radioactive materials or organic materials in high-order consumers. Although bioaccumulation is important, it is based on a dosimetric viewpoint.

The ecological system that the pale grass blue butterfly inhabits is relatively simple due to its monophagous nature [48]. Thus, this butterfly and its associated ecosystem may serve as a "model ecosystem" to investigate both the population dynamics and the environmental influences through the ecological food web after the Fukushima nuclear accident. It appears that in the case of the pale grass blue butterfly "model ecosystem," the quality of its host plant, *Oxalis corniculata,* is probably important and is determined by the quality of the soil and air. When soil is contaminated with radioactive materials and other pollutants, such as agrochemicals, the quality of the host-plant leaves decreases. Similarly, air pollutants (i.e., particulate matter) that cover the surface of leaves, whether radioactive or not, may change the physiological functions of the leaves. Thus, the quality of the soil and air will affect the health of the larval butterflies that eat the affected leaves.

The decrease in plant quality for larvae may originate from two different causes: a decrease in certain favorable chemicals (e.g., essential nutrients) in leaves and an increase in unfavorable chemicals (e.g., reactive oxygen species and defense chemicals) in leaves. In the former scenario, the lack of an essential vitamin in the leaves may be fatal for butterfly larvae because larvae are dependent on vitamins that are supplied through the ingestion of leaves. A similar case of thiamine (vitamin B_1) deficiency has been recognized as one of the major consequences of environmental pollution and destruction in Europe and North America; however, the precise causes of this deficiency are difficult to identify [77–80].

The latter possibility of the decrease in plant quality for butterfly larvae may occur if plants are stressed by even low levels of exposure to radioactive materials; this exposure can produce reactive oxygen species, defense chemicals, or another substance that is harmful to larvae. Reactive oxygen species are known to be produced by various abiotic stressors, and the production of defense chemicals are induced by insect bites in many plants [81–83]; however, whether radiation stress can trigger such responses in *O. corniculata* and in plants in general is unknown. The upregulation of unfavorable chemicals and the downregulation of favorable chemicals for larvae may occur simultaneously.

Consequently, biochemical changes in producers (i.e., plants) affect primary consumers (i.e., herbivorous animals) and then secondary consumers (i.e., carnivorous animals). These food-mediated effects of pollutants can radiate through an ecological food web, and it is indirect field effects that are different from the bioaccumulation paradigm. It is reasonable to imagine that damage to keystone species that have connections with many other species may cause relatively large effects on the ecosystem; however, recent research posits that anthropogenic disturbances on a small number of any species may cause instability in an ecosystem [84, 85].

7. Possible field effects on humans

Among the three modes of action of the field effects discussed above, the second mode (i.e., particulate matter) is associated with immunological responses that may be prominently problematic for humans because humans have very effective (and, thus, very sensitive) immunological systems, some of which insects do not have. A small amount of radioactive or nonradioactive aerosol from a nuclear reactor can potentially cause large and fatal physiological effects in some human individuals via immunological sensitization. However, immunological responses vary among individuals, and it is known that immunological sensitivity to chemicals (i.e., allergens) greatly varies among human individuals. However, once sensitized, humans can detect a remarkably small number of molecules and manifest allergic symptoms. It is possible that radioactivity denatures proteins, which makes naturally occurring proteins immunogenic. The protein-denaturing effect of ionizing radiation as well as its association with immunogenicity may be one of the important topics that should be experimentally tested. As a whole, these effects can collectively be called *the immunological effects*.

http://dx.doi.org/10.5772/intechopen.79870

The consequences of allergic reactions are complex, but one example of a type of reaction is kidney failure, which can include *nephrotic syndrome*; I have reported a case in which nephrotic syndrome was likely induced by the immunological field effects of the Fukushima nuclear accident [86]. Indeed, a general relationship between immunological sensitization and nephrotic syndrome has been demonstrated [87–92]. This relationship has not been rigorously tested; however, this is not surprising because nephrotic syndrome is a collection of diseases that have various etiologies.

Regarding the first mode of the field effect discussed above, the synergistic effects are potentially numerous in human society and in human living environments. One of the potential stressors is cedar pollen, which causes Japan-wide allergic reactions in the spring of each year, including 2011 immediately before and after the Fukushima nuclear accident. It is possible that the aerosol from the Fukushima reactors attached to cedar pollen to worsen *pollen allergy* (i.e., *hay fever*). Other potential stressors for humans may include other air pollutants, food additives, agrochemicals, and work stress. Stress resistance varies among individual humans, and some people that were not very stress resistant may have become sick after the Fukushima nuclear accident.

Regarding the third mode of the field effects discussed above, changes in plant chemicals may affect human health. Additionally, the nutritional quality of fruits and vegetables may have declined. However, different from the pale grass blue butterfly, humans are not monophagous. Moreover, vitamin supplementation is now popular in many countries including Japan. As such, this type of field effect may not manifest in humans; however, this mode may cause serious adverse impacts in the pale grass blue butterfly.

8. UNSCEAR 2017 Report

Because the field effects of "nuclear" pollution may be a new concept, at least to some extent, misunderstanding or confusion about this issue may prevail. The United Nations Scientific Committee on the Effects of Atomic Radiation (UNSCEAR) 2017 Report [93] provides an example. This report mentioned our studies in paragraph 125, in which H8 refers to Hiyama et al. [9], and M9 and M10 refer to Møller et al. [19] and Møller et al. [20], respectively.

> *125. The Committee had made reference to studies in which effects in various terrestrial biota had been observed in areas with enhanced levels of radioactive material as a result of the FDNPS accident [H8, M9, M10]. It had noted that the substantial impacts reported for populations of wild organisms from these studies were inconsistent with the main findings of the Committee's theoretical assessment. The Committee had expressed reservations about these observations, noting that uncertainties with regard to dosimetry and possible confounding factors made it difficult to substantiate firm conclusions from the cited field studies.*

It is understandable that our study is "inconsistent with the main findings of the Committee's theoretical assessment" (i.e., the dosimetric simulations). I agree that "uncertainties with regard to dosimetry" should be overcome in the near future; however, without precise dosimetric data, the findings that conclude the biological effects were correlated with the ground radiation dose and/or the distance from the nuclear reactors and that state the biological effects in the field were reproduced dose-dependently in laboratory experiments are entirely valid. The main reason for this discrepancy is the exclusion of the field effects in the UNSCEAR assessment. In contrast, our experiments were constructed to reflect real-world phenomena, including the direct effects and indirect field effects. Furthermore, contrary to the UNSCEAR statement above, there were no major confounding factors in our study [9] because it consisted of controlled laboratory experiments.

Moreover, the UNSCEAR statement completely ignores the process of logical judgment in terms of the cause of the Fukushima nuclear accident. The causality of the effects of the accident should be evaluated systematically according to logical postulates such as "the Postulates of Pollutant-Induced Biological Impacts" [45]. This includes six clauses that must be met to prove the causality of the pollutant(s) from a given source, i.e., spatial relationship, temporal relationship, direct exposure, phenotypic variability or spectrum, experimental reproduction of external exposure, and experimental reproduction of internal exposure [45]. The causality should not be judged solely from a dosimetric standpoint.

The UNSCEAR 2017 Report [91] further commented on our paper in paragraph 134, in which H9 and O12 refer to Hiyama et al. [14] and Otaki [48], respectively.

> *134. Hiyama et al. [H9] provided further evidence to suggest that the high abnormality rates observed in the pale grass blue butterfly were induced by "anthropogenic radioactive mutagens." However, Otaki [O12] synthesized the results from several studies of the effects on the same species of butterfly following the FDNPS accident, and reported that ionizing radiation was unlikely to be the exclusive source of the environmental disturbances observed.*

The above comments on our research are misleading; specifically, the last sentence wrongly implies that "the environmental disturbances observed" were caused by unknown confounding factors that were not related to the Fukushima nuclear accident. Rather, in Otaki [48], I mentioned the importance of the field effects from both radioactive and nonradioactive materials from the Fukushima Dai-ichi Nuclear Power Plant. In other words, "ionizing radiation" (i.e., the direct effects in the context of Otaki [48]) was not the exclusive source. It is entirely valid to say that the high abnormality and mortality rates observed in the butterfly were caused by the pollutants from the Fukushima nuclear accident. This UNSCEAR case indicates the low level of understanding regarding the field effects and the lack of fundamental logic among the researchers who contributed to the formulation of these paragraphs in the UNSCEAR 2017 Report [93]. On the other hand, these misleading comments may be understandable, considering that we presented the topic of indirect field effects only briefly in our previous papers. There is an urgent need for more precise explanations and experimental validation of this issue.

9. Extrapolating butterfly toxicology to humans

The evaluation of the field effects may not be straightforward because of its indirect nature; however, our system for the internal exposure experiments likely reflects both the direct effects and some of the indirect field effects based on the use of the field-collected host-plant leaves for butterfly larvae. Because the larvae are highly resistant against the internal exposure to pure radioactive cesium (unpublished data), the high mortality and abnormality rates from the contaminated leaves can be largely attributed to the indirect field effects. It should be noted that what was measured in our experiments was the radioactivity concentration of radiocesium; however, other radioactive and nonradioactive materials were released from the Fukushima nuclear reactors, and these materials may have also contaminated the leaves. In this sense, *the radioactivity concentrations of radiocesium can be considered as an indicator of the degree of the pollution*. This is an important difference from the conventional dosimetric approach. To our knowledge, quantitative toxicological data that reflected some of the field effects were available only for butterflies. Thus, it is interesting to apply these data to humans to roughly grasp the collective effects of the Fukushima nuclear accident. Although there is no rigorous reason to believe that the butterfly data are applicable to humans, this attempt can be justified because of the lack of human-specific data and data from other organisms that reflect both the direct effects and indirect field effects.

The basic experimental strategy was to collect the polluted food (i.e., plants) from Fukushima and feed the plant samples to butterfly larvae from Okinawa, which was the least polluted locality in Japan. When non-contaminated leaves were fed to larvae, normal individuals emerged. However, when polluted leaves were fed to larvae, morphologically abnormal adults emerged, and the mortality of larvae and pupae was high. The abnormality rate and the mortality rate were then obtained for each polluted diet. Because the radioactivity concentration of radiocesium species (^{134}Cs plus ^{137}Cs) in foods (Bq kg^{-1} diet) and the amount of food that each larva ate (g) was available, a dose-response curve was obtained [12].

The half abnormality dose (equivalent to median toxic dose, TD_{50}; called TD_{50} hereafter) of radiocesium for the butterfly was first obtained in Nohara et al. [10] based on the power function fit for data points from relatively high-dose diets. Later, the data points from the relatively low-dose diets were added to the previous data [11]. The mathematical model fits for these combined data were performed using the power function and Weibull function models [12]; the sigmoidal data fit with the Weibull function model yielded a TD_{50} value of 0.45 Bq body^{-1} (meaning that a cumulative dose of 0.45 Bq per larva results in abnormality or death in 50% of the population). A loose threshold was detected at approximately 10 mBq body^{-1}.

The mean body weight of larvae was 0.0346 g. Therefore, the TD_{50} can be read as 13 kBq kg^{-1} body weight. Here, I assume an average Japanese male person (30–49 years old) has a body weight of 68.5 kg, according to a survey by the Ministry of Health, Labour and Welfare [94]. For this average person, 13 kBq kg^{-1} body weight is multiplied by 68.5 kg body weight, resulting in a TD_{50} of 890.5 kBq body^{-1} for an average Japanese male human. This average person eats 1.555 kg diet day^{-1} when nutritional balance is maintained [95].

Based on these data, the radioactivity concentration of diets required to reach the TD_{50} value in a given time span in a Japanese male human can be calculated (**Figure 3a**). To consume 890.5 kBq in 1 day, 890.5 kBq must be contained in a 1.555 kg diet; thus, the radioactivity concentration of 573 kBq kg^{-1} diet must be consumed to reach the TD_{50} value in 1 day. To consume 890.5 kBq within 1 year (365 days), a 1.57 kBq kg^{-1} diet is required. Similarly, a 157 Bq kg^{-1} diet and a 15.7 Bq kg^{-1} diet are required to reach the TD_{50} value in 10 years and 100 years, respectively. Clearly, a 15.7 Bq kg^{-1} diet is mostly negligible for this average person between the age of 30 and 49 because he will naturally die before he reaches a 50% chance of becoming sick. However, *a 157 Bq kg^{-1} diet is not negligible for this average person because there is still a 50% chance of becoming sick in the next 10 years*.

Additionally, the number of days (or years) required to reach the TD_{50} value when 100 Bq kg^{-1} diet or 10 Bq kg^{-1} diet is consumed can be calculated (**Figure 3b**). *When an average Japanese male human consumes a 100 Bq kg^{-1} diet, it takes 16.7 years to reach the TD_{50} value*. This is a non-negligible time span. However, a 10 Bq kg^{-1} diet may be negligible because it takes 167 years to reach the TD_{50} value, which is beyond the human lifespan.

Considering that the amount (becquerel) of radioactivity concentration of ^{134}Cs and ^{137}Cs discussed above is as low as the amount of naturally occurring ^{40}K, a counter argument to this discussion would be that no harmful effect is expected from the conventional dosimetric view. However, it should be remembered that the amount of radiocesium is simply an indication of pollution levels in terms of the field effects. Moreover, we have experimental evidence that artificial radiocesium is clearly harmful at radioactivity levels as low as those observed for radiopotassium (unpublished data). I will discuss this important issue if there is an opportunity to do so in the future.

It should also be remembered that the discussion above completely ignored the dose-rate effects and the physiological differences between butterflies and humans, which include different biological half-lives and organ accumulation of cesium species. This study also ignored the different types of indirect field effects that may be species-specific, depending on the ecological status of a species. It should also be noted that *the TD_{50} state is toxicologically convenient to evaluate potential effects, but it means a devastating massive outbreak of diseases in terms of public health*. Another viewpoint to consider is that toxicological evaluations are often misleading and give the impression that anything that does not reach the TD_{50} value within a reasonable time or does not exceed the limit is completely safe for everybody. Scientists and politicians should pay special attention to minorities who may still be affected at this level [48, 96].

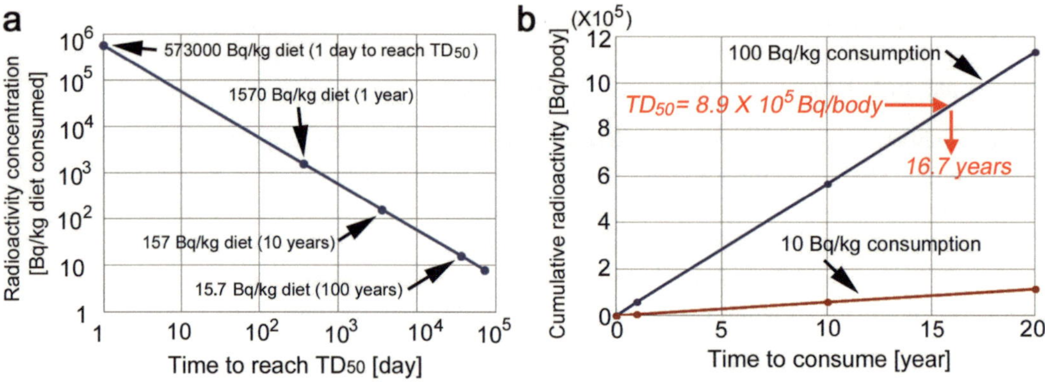

Figure 3. Extrapolation of toxicological data from the pale grass blue butterfly to an average Japanese male human. (a) Linearly extrapolating the butterfly data to understand the relationship between radioactivity concentration in consumed diet and time to reach TD_{50}. For example, to reach the TD_{50} value in 10 years, an average daily consumption of a diet containing 157 Bq kg^{-1} diet is required. (b) Linear relationship between cumulative radioactivity in a body and time to reach TD_{50}. Lines with daily 100 Bq kg^{-1} consumption and 10 Bq kg^{-1} consumption are shown. When an average of 100 Bq diet is consumed daily, it takes 16.7 years for a Japanese male human to reach the TD_{50} value (8.9×10^5 Bq body^{-1}).

Having mentioned these points, a discussion based on the TD_{50} value is probably as insightful as a discussion on the current political dose limits, which are based on the effective dose limits recommended by the ICRP [97]. In these conventional cases, no field effects were considered. Fortunately, based on the discussion above, the current regulation limit in Japan, i.e., 100 Bq kg^{-1} for general foods, may not be a completely wrong value. In fact, this value can be considered as a starting point for this type of discussion. I believe that the theoretical results above are an important first step from which we can at least present the potential values for risk assessment and management.

10. Conclusions and future perspectives

It can be concluded that the "low-dose" exposure from the Fukushima nuclear accident imposed potentially non-negligible toxic effects on organisms including butterflies and humans through field effects. At the high-dose exposure, the same field effects would exist, but they would likely be masked by the acute damage. The direct effects may be assessed reasonably by dosimetric analysis even in the field cases, especially for high-dose cases. The field-laboratory paradox is not really a paradox; rather, it indicates our fragmentary knowledge on the real-world pollution caused by this nuclear accident.

Although this chapter sheds light on one important low-dose issue, there are many other issues associated with the field effects that should be studied both in the field and in the laboratory. One of these issues is the *adaptive and evolutionary responses* of organisms to environmental radiation in contaminated areas. The pale grass blue butterfly appears to have evolutionarily adapted to the environmental pollutants [98]. This adaptive evolution may be largely in response to the field effects because the butterfly is essentially very resistant to direct irradiation without any possible adaptive response (unpublished data). However, the direct ionizing damage on DNA would also play an important role in adaptive response if such damage exists.

Simply because there are multiple effective pathways of the field effects, *sensitivity variations* to different modes may vary considerably among species and even among individuals in a given species. The net effects may be determined through synergistic amplification. To further understand the effects of the Fukushima pollution, multifaceted scientific approaches that are firmly based on field work and field-based laboratory experiments (such as the internal exposure experiments using the field-harvested leaves) are expected in the future. A mechanistic understanding of the indirect field effects is also necessary to advance this field of pollution biology.

Simultaneously, studies on the mechanisms of the direct ionizing effects in the field (although the final effects may also be affected by the indirect field effects) should be advanced. As pointed out by Steen [99], multifaceted analyses at the DNA and genomic levels are expected to reveal evidence for direct DNA damage in the field after the Fukushima nuclear accident. I believe that the immediate early exposures to short-lived radionuclides impacted DNA directly, which then might have been inherited to subsequent generations. Such evidence would firmly establish the adverse biological effects caused by the Fukushima nuclear accident at the molecular level. Furthermore, spatiotemporal changes of such DNA damage would reveal population-level dynamics of adaptive evolution in the field, serving as an important case of the real-world evolution in evolutionary biology as well as in radiation pollution biology. Borrowing the famous phrase from *Hamlet* again, I would state, "*To be and not to be* (i.e., the direct and indirect field effects), *that is the answer*".

Acknowledgements

The author thanks members of the BCPH Unit of Molecular Physiology for technical help and discussion. The author also thanks professional peer reviewers for critical comments on this manuscript. This study was thankfully funded by donators for the Fukushima Project.

Author details

Joji M. Otaki

Address all correspondence to: otaki@sci.u-ryukyu.ac.jp

The BCPH Unit of Molecular Physiology, Department of Chemistry, Biology and Marine Science, Faculty of Science, University of the Ryukyus, Okinawa, Japan

References

[1] Wheatley S, Sovacool B, Sornette D. Of disasters and dragon kings: A statistical analysis of nuclear power incidents and accidents. Risk Analysis. 2017;**37**:99-115

[2] Møller AP, Mousseau TA. Biological consequences of Chernobyl: 20 years on. Trends in Ecology and Evolution. 2006;**21**:200-207

[3] Møller AP, Mousseau TA. Strong effects of ionizing radiation from Chernobyl on mutation rates. Scientific Reports. 2015;**5**:8363

[4] Møller AP, Mousseau TA. Are organisms adapting to ionizing radiation at Chernobyl? Trends in Ecology and Evolution. 2016;**31**:281-289

[5] Aliyu AS, Evangeliou N, Mousseau TA, Wu J, Ramli AJ. An overview of current knowledge concerning the health and environmental consequences of the Fukushima Daiichi Nuclear Power Plant (FDNPP) accident. Environment International. 2015;**85**:213-228

[6] Koizumi A, Harad KH, Niisoe T, Adachi A, Fujii Y, Hitomi T, Kobayashi H, Wada Y, Watanabe T, Ishikawa H. Preliminary assessment of ecological exposure of adult residents in Fukushima Prefecture to radioactive cesium through ingestion and inhalation. Environmental Health and Preventive Medicine. 2012;**17**:292-298

[7] Strand P, Aono T, Brown JE, Garnier-Laplace J, Hosseini A, Sazykina T, Steenhuisen F, Vives i Batlle J. Assessment of Fukushima-derived radiation doses and effects on wildlife in Japan. Environmental Science and Technology Letters. 2014;**1**:198-203

[8] Fuma S, Ihara S, Takahashi H, Inaba O, Sato Y, Kubota Y, Watanabe Y, Kawaguchi I, Aono T, Soeda H, Yoshida S. Radiocaesium contamination and dose rate estimation of terrestrial and freshwater wildlife in the exclusion zone of the Fukushima Dai-ichi Nuclear Power Plant accident. Journal of Environmental Radioactivity. 2017;**171**:176-188

[9] Hiyama A, Nohara C, Kinjo S, Taira W, Gima S, Tanahara A, Otaki JM. The biological impacts of the Fukushima nuclear accident on the pale grass blue butterfly. Scientific Reports. 2012;**2**:570

[10] Nohara C, Hiyama A, Taira W, Tanahara A, Otaki JM. The biological impacts of ingested radioactive materials on the pale grass blue butterfly. Scientific Reports. 2014;**4**:4946

[11] Nohara C, Taira W, Hiyama A, Tanahara A, Takatsuji T, Otaki JM. Ingestion of radioactively contaminated diets for two generations in the pale grass blue butterfly. BMC Evolutionary Biology. 2014;**14**:193

[12] Taira W, Hiyama A, Nohara C, Sakauchi K, Otaki JM. Ingestional and transgenerational effects of the Fukushima nuclear accident on the pale grass blue butterfly. Journal of Radiation Research. 2015;**56**:i2-i8

[13] Hiyama A, Taira W, Nohara C, Iwasaki M, Kinjo S, Iwata M, Otaki JM. Spatiotemporal abnormality dynamics of the pale grass blue butterfly: Three years of monitoring (2011-2013) after the Fukushima nuclear accident. BMC Evolutionary Biology. 2015;**15**:15

[14] Hiyama A, Taira W, Iwasaki M, Sakauchi K, Gurung R, Otaki JM. Geographical distribution of morphological abnormalities and wing color pattern modifications of the pale grass blue butterfly in northeastern Japan. Entomological Science. 2017;**20**:100-110

[15] Hiyama A, Taira W, Iwasaki M, Sakauchi K, Iwata M, Otaki JM. Morphological abnormality rate of the pale grass blue butterfly *Zizeeria maha* (Lepidoptera: Lycaenidae) in southwestern Japan: A reference data set for environmental monitoring. Journal of Asia-Pacific Entomology. 2017;**20**:1333-1339

[16] Iwata M, Hiyama A, Otaki JM. System-dependent regulations of colour-pattern development: A mutagenesis study of the pale grass blue butterfly. Scientific Reports. 2013;**3**:2379

[17] Taira W, Iwasaki M, Otaki JM. Body size distributions of the pale grass blue butterfly in Japan: Size rules and the status of the Fukushima population. Scientific Reports. 2015;**5**:12351

[18] Møller AP, Mousseau TA. Low-dose radiation, scientific scrutiny, and requirements for demonstrating effects. BMC Biology. 2013;**11**:92

[19] Møller AP, Hagiwara A, Matsui S, Kasahara S, Kawatsu K, Nishiumi I, Suzuki H, Ueda K, Mousseau TA. Abundance of birds in Fukushima as judges from Chernobyl. Environmental Pollution. 2012;**164**:36-39

[20] Møller AP, Nishiumi I, Suzuki H, Ueda K, Mousseau TA. Differences in effects of radiation on abundance of animals in Fukushima and Chernobyl. Ecological Indicators. 2013; **24**:75-81

[21] Mousseau TA, Møller AP. Genetic and ecological studies of animals in Chernobyl and Fukushima. Journal of Heredity. 2014;**105**:704-709

[22] Akimoto S. Morphological abnormalities in gall-forming aphids in a radiation-contaminated area near Fukushima Daiichi: Selective impact of fallout? Ecology and Evolution. 2014;**4**:355-369

[23] Hayama S, Nakiri S, Nakanishi S, Ishii N, Uno T, Kato T, Konno F, Kawamoto Y, Tsuchida S, Ochiai K, Omi T. Concentration of radiocesium in the wild Japanese monkey (*Macaca fuscata*) over the first 15 months after the Fukushima Daiichi nuclear disaster. PLoS One. 2013;**8**:e68530

[24] Ochiai K, Hayama S, Nakiri S, Nakanishi S, Ishii N, Uno T, Kato T, Konno F, Kawamoto Y, Tsuchida S, Omi T. Low blood cell counts in wild Japanese monkeys after the Fukushima Daiichi nuclear disaster. Scientific Reports. 2014;**4**:5793

[25] Bonisoli-Alquati A, Koyama K, Tedeschi DJ, Kitamura W, Sukuzu H, Ostermiller S, Arai E, Møller AP, Mousseau TA. Abundance and genetic damage of barn swallows from Fukushima. Scientific Reports. 2015;**5**:9432

[26] Murase K, Murase J, Horie R, Endo K. Effects of the Fukushima Daiichi nuclear accident on goshawk reproduction. Scientific Reports. 2015;**5**:9405

[27] Hayashi G, Shibato J, Imanaka T, Cho K, Kubo A, Kikuchi S, Satoh K, Kimura S, Ozawa S, Fukutani S, Endo S, Ichikawa K, Agrawal GK, Shioda S, Fukumoto M, Rakwal R. Unraveling low-level gamma radiation-responsive changes in expression of early and late genes in leaves of rice seedlings at Iitate Village, Fukushima. Journal of Heredity. 2014; **105**:723-738

[28] Rakwal R, Hayashi G, Shibato J, Deepak SA, Gundimeda S, Simha U, Padmanaban A, Gupta R, Han S-I, Kim ST, Kubo A, Imanaka T, Fukumoto M, Agrawal GK, Shioda S. Progress toward rice seed OMICS in low-level gamma radiation environment in Iitate Village, Fukushima. Journal of Heredity. 2018;**109**:206-211

[29] Watanabe Y, Ichikawa S, Kubota M, Hoshino J, Kubota Y, Maruyama K, Fuma S, Kawaguchi I, Yoschenko VI, Yoshida S. Morphological defects in native Japanese fir trees around the Fukushima Daiichi Nuclear Power Plant. Scientific Reports. 2015;**5**:13232

[30] Yoschenko V, Nanba K, Yoshida S, Watanabe Y, Takase T, Sato N, Keitoku K. Morphological abnormalities in Japanese red pine (*Pinus densiflora*) at the territories contaminated as a result of the accident at Fukushima Dai-ichi Nuclear Power Plant. Journal of Environmental Radioactivity. 2016;**165**:60-67

[31] Horiguchi T, Yoshii H, Mizuno S, Shiraishi H. Decline in intertidal biota after the 2011 Great East Japan Earthquake and Tsunami and the Fukushima nuclear disaster: Field observations. Scientific Reports. 2016;**6**:20416

[32] Urushihara Y, Kawasumi K, Endo S, Tanaka K, Hirakawa Y, Hayashi G, Sekine T, Kino Y, Kuwahara Y, Suzuki M, Fukumoto M, Yamashiro H, Abe Y, Fukuda T, Shinoda H, Isogai E, Arai T, Fukumoto M. Analysis of plasma protein concentrations and enzyme activities

in cattle within the ex-evacuation zone of the Fukushima Daiichi nuclear plant accident. PLoS One. 2016;**11**:e0155069

[33] Morimoto M, Kato A, Kobayashi J, Okuda K, Kuwahara Y, Kino Y, Abe Y, Sekine T, Fukuda T, Isogai E, Fukumoto M. Gene expression analyses of the small intestine of pigs in the ex-evacuation zone of the Fukushima Daiichi Nuclear Power Plant. BMC Veterinary Research. 2017;**13**:337

[34] Nakamura AJ, Suzuki M, Redon CE, Kuwahara Y, Yamashiro H, Abe Y, Takahashi S, Fukuda T, Isogai E, Bonner WM, Fukumoto M. The causal relationship between DNA damage induction in bovine lymphocytes and the Fukushima Nuclear Power Plant Accident. Radiation Research. 2017;**187**:630-636

[35] Takino S, Yamashiro H, Sugano Y, Fujishima Y, Nakata A, Kasai K, Hayashi G, Urushihara Y, Suzuki M, Shinoda H, Miura T, Fukumoto M. Analysis of the effect of chronic and low-dose radiation exposure on spermatogenic cells of male large Japanese field mice (*Apodemus speciosus*) after the Fukushima Daiichi Nuclear Power Plant Accident. Radiation Research. 2017;**187**:161-168

[36] Kubota Y, Tsuji H, Kawagoshi T, Shiomi N, Takahashi H, Watanabe Y, Fuma S, Doi K, Kawaguchi I, Aoki M, Kubota M, Furuhara Y, Shigemura Y, Mizoguchi M, Yamada F, Tomozawa M, Sakamoto SH, Yoshida S. Chromosomal aberrations in wild mice captured in areas differentially contaminated by the Fukushima Dai-ichi Nuclear power plant accident. Environmental Science and Technology. 2015;**49**:10074-10083

[37] Kawagoshi T, Shiomi N, Takahashi H, Watanabe Y, Fuma S, Doi K, Kawaguchi I, Aoki M, Kubota M, Furuhara Y, Shigemura Y, Mizoguchi M, Yamada F, Tomozawa M, Sakamoto SH, Yoshida S, Kubota Y. Chromosomal aberrations in large Japanese field mice (*Apodemus speciosus*) captured near Fukushima Dai-ichi nuclear power plant. Environmental Science and Technology. 2017;**51**:4632-4641

[38] Yamashiro H, Abe Y, Fukuda T, Kino Y, Kawaguchi I, Kuwahara Y, Fukumoto M, Takahashi S, Suzuki M, Kobayashi J, Uematsu E, Tong B, Yamada T, Yoshida S, Sato E, Shinoda H, Sekine T, Isogai E, Fukumoto M. Effects of radioactive caesium on bull testes after the Fukushima nuclear accident. Scientific Reports. 2013;**3**:2850

[39] Yamashiro H, Abe Y, Hayashi G, Urushihara Y, Kuwahara Y, Suzuki M, Kobayashi J, Kino Y, Fukuda Y, Tong B, Takino S, Sugano Y, Sugimura S, Yamada T, Isogai E, Fukumoto M. Electron probe X-ray microanalysis of boar and inobuta testes after the Fukushima accident. Journal of Radiation Research. 2015;**56**:i42-i47

[40] Okano T, Ishiniwa H, Onuma M, Shindo J, Yokohata Y, Tamaoki M. Effects of environmental radiation on testes and spermatogenesis in wild large Japanese field mice (*Apodemus speciosus*) from Fukushima. Scientific Reports. 2016;**6**:23601

[41] Nakajima H, Yamaguchi Y, Yoshimura T, Fukumoto M, Todo T. Fukushima simulation experiment: Assessing the effects of chronic low-dose-rate internal ^{137}Cs radiation exposure on litter size, sex ratio, and biokinetics in mice. Journal of Radiation Research. 2015;**56**:i29-i35

[42] Akimoto S, Li Y, Imanaka T, Sato H, Ishida K. Effects of radiation from contaminated soil and moss in Fukushima on embryogenesis and egg hatching of the aphid *Prociphilus oriens*. Journal of Heredity. 2018;**109**:199-205

[43] Garnier-Laplace J, Geras'kin S, Delta-Vedova C, Beaugelin-Seiller K, Hinton TG, Real A, Oudalova A. Are radiosensitivity data derived from natural field conditions consistent with data from controlled exposures? A case study of Chernobyl wildlife chronically exposed to low dose rates. Journal of Environmental Radioactivity. 2013;**121**:12-21

[44] Hiyama A, Nohara C, Taira W, Kinjo S, Iwata M, Otaki JM. The Fukushima nuclear accident and the pale grass blue butterfly: Evaluating biological effects of long-term low-dose exposures. BMC Evolutionary Biology. 2013;**13**:168

[45] Taira W, Nohara C, Hiyama A, Otaki JM. Fukushima's biological impacts: The case of the pale grass blue butterfly. Journal of Heredity. 2014;**105**:710-722

[46] Lawrence PA. The Making of a Fly: The Genetics of Animal Design. Oxford: Blackwell Scientific; 1992

[47] Caro T. Conservation by Proxy: Indicator, Umbrella, Keystone, Flagship, and Other Surrogate Species. 2nd ed. Washington: Island Press; 2010

[48] Otaki JM. Fukushima's lessons from the blue butterfly: A risk assessment of the human living environment in the post-Fukushima era. Integrated Environmental Assessment and Management. 2016;**12**:667-672

[49] Hiyama A, Iwata M, Otaki JM. Rearing the pale grass blue *Zizeeria maha* (Lepidoptera, Lycaenidae): Toward the establishment of a lycaenid model system for butterfly physiology and genetics. Entomological Science. 2010;**13**:293-302

[50] Otaki JM, Hiyama A, Iwata M, Kudo T. Phenotypic plasticity in the range-margin population of the lycaenid butterfly *Zizeeria maha*. BMC Evolutionary Biology. 2010;**10**:252

[51] UNSCEAR. United Nations Scientific Committee on the Effects of Atomic Radiation. Biological Mechanisms of Radiation Actions at Low Doses. A White Paper to Guide the Scientific Committee's Future Programme of Work. New York: United Nations; 2012

[52] Mothersill C, Seymour C. Implications for environmental health of multiple stressors. Journal of Radiological Protection. 2009;**29**:A21

[53] Mothersill C, Seymour C. Implications for human and environmental health of low doses of ionizing radiation. Journal of Environmental Radioactivity. 2014;**133**:5-9

[54] Mothersill C, Rusin A, Fernandez-Palomo C, Seymour C. History of bystander effects research 1905-present; what is in a name? International Journal of Radiation Biology. 2017;**29**:1-12

[55] Mothersill C, Rusin A, Seymour C. Low doses and non-targeted effects in environmental radiation protection; where are we now and where should we go? Environmental Research. 2017;**159**:484-490

[56] Mothersill C, Seymour C. Old data-new concepts: Integrating "indirect effects" into radiation protection. Health Physics. 2018;**115**:170-178

[57] Otaki JM, Taira W. Current status of the blue butterfly in Fukushima research. Journal of Heredity. 2018;**109**:178-187

[58] Leenhouts HP, Chadwick KH. An analysis of synergistic sensitization. British Journal of Cancer. 1978;**3**:S198-S201

[59] Borek C, Zaider M, Ong A, Mason H, Witz G. Ozone acts alone and synergistically with ionizing radiation to induce *in vitro* neoplastic transformation. Carcinogenesis. 1986;**7**:1611-1613

[60] Mothersill C, Salbu B, Heier LS, Teien HC, Denbeigh J, Oughton D, Rosseland BO, Seymour CB. Multiple stressor effects of radiation and metals in salmon (*Salmo salar*). Journal of Environmental Radioactivity. 2007;**96**:20-31

[61] Manti L, D'Arco A. Cooperative biological effects between ionizing radiation and other physical and chemical agents. Mutation Research. 2010;**704**:115-122

[62] Gilbert SF, Epel D. Ecological Developmental Biology: The Environmental Regulation of Development, Health, and Evolution. 2nd ed. Sunderland: Sinauer Associates; 2015

[63] Relyea RA. Predator cues and pesticides: A double dose of danger for amphibians. Ecological Applications. 2003;**13**:1515-1521

[64] Relyea RA. Fine-tuned phenotypes: Tadpole plasticity under 16 combinations of predators and competitors. Ecology. 2004;**85**:172-179

[65] Ballard JD. Pathogens boosted by food additive. Nature. 2018;**553**:285-286

[66] Collins J, Robinson C, Danhof H, Knetsch CW, van Leeuwen HC, Lawley TD, Auchtung JM, Britton RA. Dietary trehalose enhances virulence of epidemic *Clostridium difficile*. Nature. 2018;**553**:291-294

[67] Salbu B, Lind OC. Radioactive particles released to the environment from the Fukushima reactors—Confirmation is still needed. Integrated Environmental Assessment and Management. 2016;**12**:687-689

[68] Kaltofen M, Gundersen A. Radioactively-hot particles detected in dusts and soils from Northern Japan by combination of gamma spectrometry, autoradiography, and SEM/EDS analysis and implications in radiation risk assessment. Science of Total Environment. 2017;**607-608**:1065-1072

[69] Seaton A, Godden D, MacNee W, Donaldson K. Particulate air pollution and acute health effects. Lancet. 1995;**345**:176-178

[70] Kappos A, Bruckmann P, Eikmann T, Englert N, Heinrich U, et al. Health effects of particles in ambient air. International Journal of Hygiene and Environmental Health. 2004;**207**:399-407

[71] Utell MJ, Frampton MW. Acute health effects of ambient air pollution: The ultrafine particle hypothesis. Journal of Aerosol Medicine. 2009;**13**:355-359

[72] Shiraiwa M, Selzle K, Pöschl U. Hazardous components and health effects of atmospheric aerosol particles: Reactive oxygen species, soot, polycyclic aromatic compounds and allergenic proteins. Free Radical Research. 2012;**46**:927-939

[73] UNSCEAR. United Nations Scientific Committee on the Effects of Atomic Radiation. Ionizing Radiation: Sources and Biological Effects. 1982 Report to the General Assembly, with Annexes. New York: United Nations; 1982

[74] Adachi K, Kajino M, Zaizen Y, Igarashi Y. Emission of spherical cesium-bearing particles from an early stage of the Fukushima nuclear accident. Scientific Reports. 2013;**3**:2554

[75] Miyamoto Y, Yasuda K, Magara M. Size distribution of radioactive particles collected at Tokai, Japan 6 days after the nuclear accident. Journal of Environmental Radioactivity. 2014;**132**:1-7

[76] Bréchignac F, Oughton D, Mays C, Barnthouse L, Beasley JC, Bonisili-Alquati A, Bradshaw C, Brown J, Dray S, Geras'kin S, Glenn T, Higley K, Ishida K, Kapustka L, Kautsky U, Kuhne W, Lynch M, Mappes T, Mihok S, Møller AP, Mothersill C, Mousseau TA, Otaki JM, Pryakhin E, Rhodes OE Jr, Salbu B, Strand P, Tsukada H. Addressing ecological effects of radiation on populations and ecosystems to improve protection of the environment against radiation: Agreed statements from a Consensus Symposium. Journal of Environmental Radioactivity. 2016;**158-159**:21-29

[77] Sañudo-Wilhelmy SA, Cutter LS, Durazo R, Smail EA, Gómez-Consamau L, Webb EA, Prokopenko MG, Berelson WM, Karl DM. Multiple B-vitamin depletion in large areas of the coastal ocean. Proceeding of the National Academy of Sciences of the United States of America. 2012;**109**:14041-14045

[78] Balk L, Hägerroth P-Å, Åkerman G, Hanson M, Tjärnlund U, Hansson T, Hallgrimsson GT, Zebühr Y, Broman D, Mörner T, Sundberg H. Wild birds of declining European species are dying from a thiamine deficiency syndrome. Proceeding of the National Academy of Sciences of the United States of America. 2009;**106**:12001-12006

[79] Balk L, Hägerroth P-Å, Gustavsson H, Sigg L, Åkerman G, Muñoz YR, Honeyfield DC, Tjärnlund U, Oliveira K, Ström K, McCormick SD, Karlsson S, Ström M, van Manen M, Berg A-L, Halldórsson HP, Strömquist J, Collier TK, Börjeson H, Mörner T, Hansson T. Widespread episodic thiamine deficiency in Northern Hemisphere wildlife. Scientific Reports. 2016;**6**:38821

[80] Sonne C, Alstrup O, Therkildsen OR. A review of the factors causing paralysis in wild birds: Implications for the paralytic syndrome observed in the Baltic Sea. Science of Total Environment. 2012;**416**:32-39

[81] Dicke M, van Poecke RMP. Signalling in plant-insect interactions: Signal transduction in direct and indirect plant defense. In: Scheel D, Wasternack C, editors. Plant Signal Transduction. Oxford: Oxford University Press; 2002. pp. 289-316

[82] Kessler A, Baldwin IT. Plant responses to insect herbivory: The emerging molecular analysis. Annual Review of Plant Biology. 2002;**53**:299-328

[83] Taiz L, Zeiger E, Møller IM, Murphy A. Plant Physiology and Development. 6th ed. Sunderland: Sinauer Associates; 2015

[84] McCann KS. The diversity–stability debate. Nature. 2000;**405**:228-233

[85] Ives AR, Carpenter SR. Stability and diversity of ecosystems. Science. 2007;**317**:58-62

[86] Otaki JM. Fukushima nuclear accident: Potential health effects inferred from butterfly and human cases. In: D'Mello JPF, editor. Environmental Toxicology. Oxon: CABI Publishing; 2019 In press

[87] Pirotzky E, Hieblot C, Benveniste J, Laurent J, Lagrue G, Noirot C. Basophil sensitisation in idiopathic nephrotic syndrome. Lancet. 1982;**319**:358-361

[88] Yap HK, Yip WC, Lee BW, Ho TF, Teo J, Aw SE, Tay JS. The incidence of atopy in steroid-responsive nephrotic syndrome: Clinical and immunological parameters. Annals of Allergy. 1983;**51**:590-594

[89] Laurent J, Lagrue G, Belghiti D, Noirot C, Hirbec G. Is house dust allergen a possible causal factor for relapses in lipoid nephrosis? Allergy. 1984;**39**:231-236

[90] Lin CY, Lee BH, Lin CC, Chen WP. A study of the relationship between childhood nephrotic syndrome and allergic diseases. Chest. 1990;**97**:1408-1411

[91] Abdel-Hafez M, Shimada M, Lee PY, Johnson RJ, Garin EH. Idiopathic nephrotic syndrome and atopy: Is there a common link? American Journal of Kidney Diseases. 2009;**54**:945-954

[92] Cohen EP, Fish BL, Moulder JE. Late-onset effects of radiation and chronic kidney diseases. Lancet. 2015;**386**:1737-1738

[93] UNSCEAR. United Nations Scientific Committee on the Effects of Atomic Radiation. Developments since the 2013 UNSCEAR Report on the Levels and Effects of Radiation Exposure due to the Nuclear Accident following the Great East-Japan Earthquake and Tsunami. A 2017 White Paper to Guide the Scientific Committee's Future Programme of Work. New York: United Nations; 2017

[94] JATCC. Japan Association of Training Colleges for Cooks. Characteristics of Foods and Nutrients. Ministry of Health, Labour and Welfare: Tokyo; 2014

[95] Kagawa Y. Standard Tables of Food Consumption in Japan. 7th ed. Sakado: Kagawa Nutrition University Press; 2016

[96] Fukunaga H, Yokoya A. Low-dose radiation risk and individual variation in radiation sensitivity in Fukushima. Journal of Radiation Research. 2016;**57**:98-100

[97] Iwaoka K. The current limits for radionuclides in food in Japan. Health Physics. 2016;**111**:471-478

[98] Nohara C, Hiyama A, Taira W, Otaki JM. Robustness and radiation resistance of the pale grass blue butterfly from radioactively contaminated areas: A possible case of adaptive evolution. Journal of Heredity. 2018;**109**:188-198

[99] Steen TY. Ecological impacts of ionizing radiation: Follow-up studies of nonhuman species at Fukushima. Journal of Heredity. 2018;**109**:176-177

Chapter 5

Integrated Policymaking for Realizing Benefits and Mitigating Secondary Impacts of Cold Fusion

Thomas W. Grimshaw

Additional information is available at the end of the chapter

http://dx.doi.org/10.5772/intechopen.78323

Abstract

The potential benefits of LENR as an energy source have been well understood since its announcement in 1989. Improved prospects of LENR in recent years are indicated by the significant numbers and varied locations of researchers in several countries, a large body of accumulated evidence, advances in development of explanations, and favorable LENR device developments. The changing landscape creates policymaking opportunities for supporting LENR to realize its benefits, planning proactively to deal with anticipated impacts, and integrating the updates as a comprehensive policy program. Policy updates for LENR support may be accomplished in an evidence-based policymaking framework. The level of evidence for LENR indicates that updates should include at least research comparable to other emerging energy technologies. Broad LENR deployment for energy supply is expected to have major secondary impacts as a disruptive technology. Technology assessment is a readily available methodology for developing mitigative measures. The public interest will be served by integrating LENR policies for its development and impact mitigation. For example, policies for secondary impacts can be formulated based on LENR support policies and the pace of its deployment. Updated policies may also be integrated at the national and international level and between the public and private sectors.

Keywords: cold fusion, LENR, energy policy, evidence-based policymaking, secondary impacts, disruptive energy technology, technology assessment

1. Introduction

Cold fusion (widely referred to as low energy nuclear reactions LENR) presents major opportunities to enhance the public interest as a potential new source of cheap and clean energy. Although LENR was rejected by mainstream science within a year or so of its announcement

in March 1989, the phenomenon has continued to be extensively researched. LENR's improved prospects in recent years have resulted in a need for updates in LENR policies. Policymaking opportunities are emerging in three main areas—supporting LENR to realize its potential benefits, planning proactively to deal with its anticipated adverse secondary impacts, and integrating the updates in a comprehensive policy program. The objectives of this paper are to:

- Review the changing landscape of LENR
- Describe opportunities for updating policies for support of LENR development
- Delineate potential policies for mitigating adverse secondary impacts
- Analyze opportunities for integrating LENR policies both nationally and internationally
- Summarize the benefits and challenges of achieving updated and integrated policies

The world is in desperate need of new sources of clean and inexpensive energy. If this were not the case, cold fusion would perhaps be just a curiosity in the history of science.

2. The changing landscape of LENR

Improved LENR prospects are indicated by at least four lines of argument—the significant numbers and varied locations of researchers in several countries, the resulting large body of accumulated evidence, advances in development of explanations, and recent favorable events.

2.1. Continued research worldwide

Unlike most claims of new phenomena that are not accepted by mainstream science, LENR research was not discontinued after it was rejected. On the contrary, many investigators have continued to work in the field, resulting in a large body of evidence for LENR reality. For example, at least 50 investigators in nine countries (including the U.S., Italy, Japan, India, Russia, and China) have continued LENR research. An international LENR society (International Society of Condensed Matter Nuclear Science, ISCMNS) was formed several years ago [1], and an affiliated journal dedicated to LENR research reporting (Journal of ISCMNS) is published online quarterly.

International conferences are held in countries around the world every one to 2 years, with a typical attendance of about 200. Twenty conferences (International Conferences on Cold Fusion, ICCFs) have been held since they were begun in 1990. Attendance at the 2015 conference (ICCF-19), which took place in Italy, was nearly 600. The 2016 conference (ICCF-20) was in Sendai, Japan, and the 2018 conference (ICCF-21) is planned for Fort Collins, Colorado (campus of Colorado State University) in 2018 [2]. A substantial community of LENR researchers and other interested parties has emerged. Its size is indicated by the CMNS Google Group, which was formed over 10 years ago and currently has over 300 participants.

Although the U.S. Department of Energy has not provided leadership in LENR research, investigations continued at several other U.S. agencies after the 1989 rejection. For example, the U.S. National Aeronautics and Space Administration (NASA) has conducted research at both the Glenn and Langley research centers [3, 4]. Elements of the U.S. Department of Defense (DoD) have also continued research and related interests. The Defense Intelligence Agency (DIA) assessed "with high confidence that if LENR can produce nuclear-origin energy at room temperatures, this disruptive technology could revolutionize energy production and storage, since nuclear reactions release millions of times more energy per unit mass then do (sic) any known chemical fuel" [5].

Several components of the U.S. Navy have also had active LENR research efforts. The U.S. Naval Research Laboratory (NRL), for example, worked on LENR beginning when the field started in 1989. Other Navy organizations have also pursued LENR research and related activities, including the U.S. Space and Naval Warfare Systems Command (SPAWAR), U.S. Naval Air Weapons Station (China Lake), and the U.S. Naval Postgraduate School.

An industrial association (LENRIA, for LENR Industrial Association) was formed in about 2013 to promote LENR development. LENRIA seeks to "advocate for both scientific study and, especially, commercial advancement of the field" [6]. It envisions a LENR ecosystem consisting of more than 30 R&D concerns, government entities, corporations, private labs, and publications and websites. LENRIA is sponsoring ICCF-21 in June 2018. In early 2017, The Anthropocene Institute published a report that included a list of almost 100 LENR-related entities (another "LENR Ecosystem") in five categories [7]: Makers (37); R&D Organizations (41); Investment Funds (7); Commercial Equipment Suppliers (5); and Non-Profits (6).

2.2. Large and growing body of evidence

The substantial research in LENR has resulted in a large accumulation of evidence for its reality. One indicator of this evidence is a website dedicated to collecting LENR publications (LENR-CANR.org), which has a bibliography of more than 3800 journal papers and related items. As of March 2018, about 4.6 million visits had been made and more than 4.2 million papers had been downloaded [8] from the website.

Storms [9] has documented 380 papers reporting LENR just up to about 2007 as indicated by four signatures—anomalous heat (184 reports), tritium (61), transmutation (80), and radiation (55). Many more reports have been prepared in the subsequent years. Storms and Grimshaw [10] examined the evidence for LENR in relation to published criteria for distinguishing science from pseudoscience by Langmuir [11], Sagan [12], and Shermer [13]. Twenty-seven criteria were compiled, and LENR was examined in relation to each criterion. It was found that the criteria were satisfied, and it was concluded that LENR research is science and not pseudoscience.

2.3. Advances in theory development

Significant progress has also been made in developing an explanation of LENR. Many hypotheses have been advanced, but much remains to be done to converge on a full explanation. Two well-known examples are the hypotheses advanced by Peter Hagelstein of MIT and Edmund Storms, who is retired from Los Alamos National Laboratory.

Hagelstein [14] notes that LENR is indicated by the large amount of energy produced, the absence of expected chemical products, and the presence of expected amounts of new helium-4 in palladium deuteride experiments where LENR is observed. He observes that there appears to be no other conclusion besides a nuclear origin for the observations, but that there is a lack in LENR of the usual radiation signals that are used to study nuclear reactions. Hagelstein's hypothesis includes both conventional and new physics. In palladium deuteride systems reactions occur in vacancies in the lattice. The reactions involve fractionation of a large nuclear quantum combined with a coupling mechanism involving vibration and nuclei. Hagelstein utilizes the fundamental relativistic Hamiltonian in the explanation. The approach thus uses new concepts on a foundation of established physics.

Storms' hypothesis [15] proposes small sites, termed "nuclear active environments" (NAEs), that are located at or close to the surface rather than in the lattice, as is postulated by Hagelstein. These NAEs form in microcracks that are typically caused by stress relief in the material. Hydrogen atoms migrate into the NAEs and form linear structures called "hydrotons." Vibration of the atoms in the hydroton results in nuclear reactions, with release of energy as photons that are absorbed in the lattice. The mechanism of the nuclear reactions in the hydroton has not yet been explained but would almost certainly involve new physics.

2.4. Developments in recent years

The case for LENR is strengthened by several occurrences in the field in the last few years. One of the most significant of these was the emergence of research centers at several universities. The Sidney Kimmel Institute for Nuclear Renaissance (SKINR) was formed at the University of Missouri in 2012 to perform fundamental research aimed at discovery of the mechanisms of the anomalous heat effect (AHE), a term used for LENR. Experiments are performed in four areas—nuclear mechanism, general mechanism, solid state theory, and cathode development (for electrolytic cells) [16]. The Center for Emerging Energy Science (CEES) was founded at Texas Tech University in 2015 to explore critical parameters in the observation of AHE [17]. The intent of its work is to gain fundamental understanding of the LENR mechanisms. A Condensed Matter Nuclear Reactions Division was also recently formed at Tohoku University in Sendai, Japan. Three purposes have been advanced for the Division—fundamental LENR research, development of a new energy generation method, and determination of a new approach for nuclear waste decontamination [18]. This organization sponsored ICCF-20 in Sendai in October 2016.

Further indication that cold fusion potential may be realized is the significant number of LENR-based devices that have been introduced in recent years. One major example is Andrea Rossi's E-Cat (for "energy catalyzer"), which is based on a nickel-hydrogen setup. Several demonstrations of this device were held in 2011, culminating in a multiple-unit test in October 2011. About 2350 kWh of energy was reported for this test [19]. A three-part test of a high-temperature version of Rossi's device (E-Cat HT or "Hot Cat") was subsequently performed [20]. The first part of the test was not considered successful because the reactor melted before meaningful data could be obtained. The second test reportedly produced 195 kWh of energy. The third part was indicated to produce 95 kWh.

Another set of experiments, consisting of two phases, was subsequently performed with a different E-Cat design [21]. These experiments are frequently referred to as the "Lugano test" for the location in Switzerland where they were performed. During the 32-day test, 1.4 MWh of net energy was reported. The experiments also included analyses of the isotopes of in the energy-producing contents of the E-Cat. Observed shifts in the isotope composition before and after the tests were inferred to be the result of nuclear reactions. The large amounts of energy produced, high ratios of output to input power, and changes in isotope content were interpreted as evidence of LENR. It was announced in 2014 that the firm Industrial Heat (IH) had acquired partial rights to Rossi's E-Cat technology [22]. However, the relationship between Rossi and IH did not have a positive outcome and became litigious. A lawsuit between the entities was settled in 2016.

Investigation of devices apparently similar in design to the E-Cat has continued, notably in Russia and China. Parkhomov [23, 24], a retired researcher from Lomonosov Moscow State University, reported experiments performed with two different but related designs. Both devices were configured to approximate the Lugano test of Rossi's E-Cat, but with significant differences, including the method of heat measurement. A principal conclusion was that the devices, described as "similar to (the) high-temperature Rossi heat generator … produce more energy than they consume" above temperatures above about 1100° C. It was also concluded that the second device produced more than 40 kWh of excess energy.

Jiang is a retired researcher affiliated with the Ni-H Research Group at the China Institute of Atomic Energy in Beijing. His reactor design and materials are somewhat similar to those of Parkhomov with a setup approximating the Lugano test [25]. The experiment was apparently performed for over 12 hours, during which 600 W of excess heat was observed for a portion of that time. The reported ratio of the 600 W to the input power of 780 W was 0.77. Jiang concluded that "the origin of excess heat cannot be explained by chemical energy."

JET Energy, Inc. has conducted LENR research with two types of devices called the NANOR and PHUSOR [26], both of which utilize deuterium. The PHUSOR is an aqueous configuration that uses palladium or nickel with the deuterium. The NANOR is non-aqueous and uses nanoscale particles consisting mainly of palladium, zirconium, and nickel. JET Energy maintains close collaboration with the Energy Production and Conversion Group at MIT [27].

Brillouin Energy [28] has developed LENR-based technology for energy production using hydrogen and nickel (or other metal with appropriate properties). The technology is referred to as "Controlled Electron Capture Reaction" (CECR). Hydrogen is brought into contact with nickel, and reactions are stimulated with electromagnetic pulses. The energy is reported to be in the form of heat that is absorbed by the metal and captured for beneficial use. An apparently updated version of the Brillouin approach and technology (HHT™) was recently reported [29].

Although none of these examples has demonstrated a working device having practical applications or commercial production, when considered in aggregate they provide further evidence that LENR may yet fulfill its potential as a source of energy. Overall, a changing landscape for LENR is indicated by the substantial number of researchers, the accumulated body of research, and progress in developing theories. The recent emergence of academic research entities and the proposed LENR energy devices also seem to strengthen the cold fusion case.

Three goals must be achieved for LENR and its benefits to be realized—more consistent reproducibility, fuller explanation of the process, and demonstration of its ability to produce usable amounts of energy. These goals may be achieved with affirmative policies for increased R&D support.

3. Policy updates for LENR support

The first policymaking opportunity resulting from LENR's changing landscape is revision of current policies for LENR support. Updates in these policies may best be accomplished in a framework of evidence-based policymaking (EBP) [30, 31]. The policy options (PO) are:

1. Discontinue research entirely (unlikely given the continuing interest)
2. Business as usual—continued marginalization
3. Reinstatement and development with other emerging energy technologies
4. Enhanced support, perhaps on a par with hot fusion
5. Crash program, possibly like the Manhattan Project during World War II, to realize LENR's benefits.

Selecting the alternative that best serves the public interest may be challenging because of the history and continuing rejection of LENR. Policymaking is further complicated by a need for improved reproducibility and a better explanation of the LENR phenomenon. To deal with these complications, LENR policy may be analyzed and established in terms of level of evidence (LOE) for its existence:

1. Preponderance of evidence (>50% probability)
2. Clear and convincing evidence (>70%)
3. Beyond a reasonable doubt (>90%)

The LOE may be further interpreted for decisions on appropriate policy responses. At least a preponderance of evidence may reasonably be inferred from the large number researchers, the major body of evidence that has been accumulated, and the progress in achieving LENR explanation. Clear and convincing evidence is indicated by the emergence of LENR-dedicated research centers at several universities and by the significant number of proposed devices that purport to produce energy from LENR. When sufficient reproducibility and an adequate explanation are achieved, it may be asserted that the evidence is sufficient to demonstrate LENR beyond a reasonable doubt.

Policy responses to these proposed levels of evidence may also be suggested. If LENR is indicated with a preponderance of evidence, it should be fully reinstated and pursued with other emerging energy technologies. If there is clear and convincing evidence, a higher level

http://dx.doi.org/10.5772/intechopen.78323

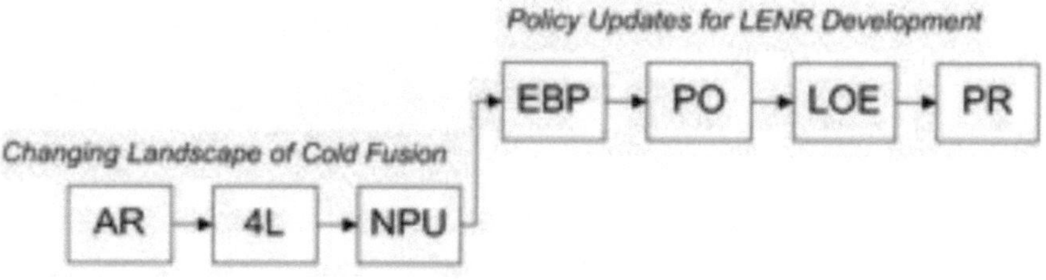

Figure 1. The changing landscape of LENR and resulting need for policy updates for its support. AR—Announcement and rejection (1989); 4 L—Four lines of argument; NPU—Need for policy updates; EBP—Evidence-based policymaking framework; PO—Five policy options; LOE—Level of evidence for LENR; PR—Policy responses (updates).

of support is needed, perhaps comparable to hot fusion support over the past five decades. If LENR is indicated beyond a reasonable doubt, it may be appropriate to institute a crash program similar to the Manhattan Project, which resulted in the atomic bomb in World War II.

In summary, it appears based on the level of evidence that LENR should at a minimum be reinstated and researched fully. It may in fact warrant investigation and development at a level similar to hot fusion research. **Figure 1** shows diagrammatically how the changing landscape of LENR leads to the need for policy updates for its support. The changing landscape began sometime after LENR's 1989 announcement and rejection (AR). The four lines of argument (4 L) for its improved prospects described above lead to a need for policy updates (NPU). The updates are founded on evidence-based policymaking (EBP). The five policy options (PO) are evaluated by the level of evidence (LOE) for LENR existence, leading to the appropriate policy responses (PR)—reinstate and research fully or provide more enhanced support.

4. Policies for mitigating adverse secondary impacts

The second policymaking opportunity resulting from LENR's changing landscape is to address potential adverse secondary impacts with proactive planning. Broad deployment of LENR for energy supply may be expected to have major secondary impacts as a disruptive technology [32, 33]. Direct impacts are anticipated for all phases of the energy chain—supply, transport, storage, and consumption. Indirect impacts will be felt most by the components of society that are closely tied to the energy cycle, such as the affected sectors of the workforce and the communities that rely on energy activities (e.g., coal mining towns).

Technology Assessment (TA) is a mature and well-established method for addressing both direct and indirect secondary impacts and may readily be applied to cold fusion case [34, 35]. The stages of a TA application are generally as follows:

1. Identify impacts
2. Determined affected parties
3. Develop mitigation strategy

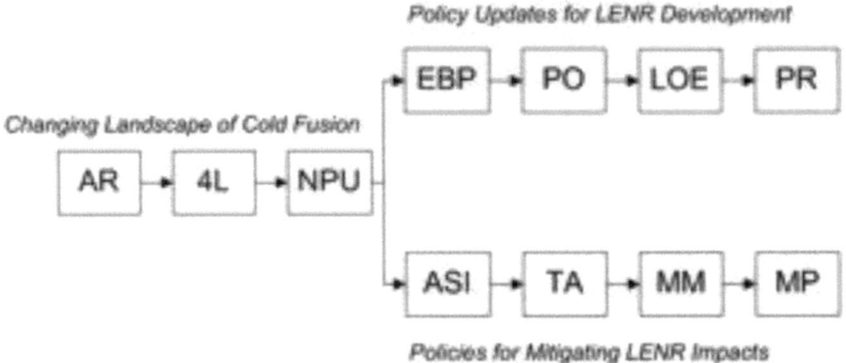

Figure 2. Illustration of need for policies to mitigate adverse secondary impacts resulting from the changing LENR landscape. ASI—Adverse secondary impacts; TA—Technology assessment methodology; MM—Mitigating measures; MP—Overall mitigation plan.

4. Define sources of assistance (e.g., agencies)
5. Engage representatives (e.g., advisory group)
6. Define mitigation measures for both direct and indirect impacts
7. Develop and implement mitigation plan

TA enables proactive planning to mitigate impacts and has ample precedent for application to energy-related issues [36, 37]. **Figure 2** summarizes how the changing LENR landscape leads to the need for policies for mitigating adverse secondary impacts in addition to required policy updates for supporting LENR development. Adverse secondary impacts (ASI) stem from the need for policy updates (NPU) and are addressed by technology assessment methodology (TA). Mitigating measures (MM) are defined, leading to an overall mitigation plan (MP).

5. Opportunities for integrating LENR policies

A third policymaking opportunity for LENR is to integrate the policy actions and updates. For example, policies for mitigation planning for secondary impacts can be coordinated with the pace of LENR development and deployment. Policies can also be integrated among agencies at the national level, between the public and private sectors, and among nations.

5.1. Integration of mitigation planning with LENR development support

As LENR prospects improve as a result of increased support, mitigation planning can be adjusted for the changing imminence and rate of deployment. This adjustment would be necessary to achieve the objectives of proactive planning for mitigation. **Figure 3** shows how the policy response for LENR development (PR) and resulting rate of deployment guides planning for mitigation (GP) as the overall plan (MP) is prepared.

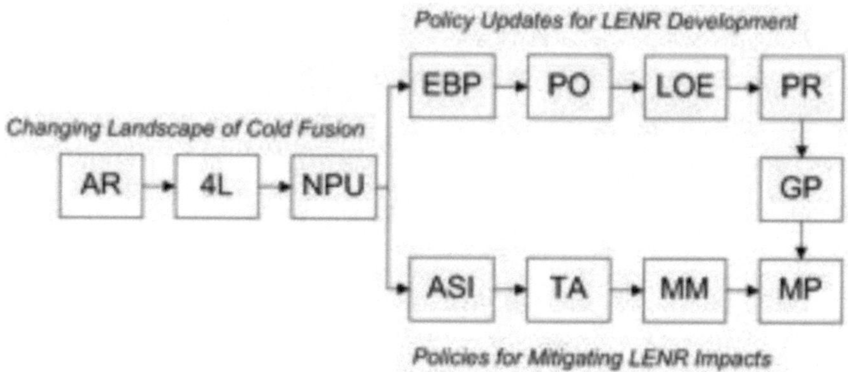

Figure 3. The pace of mitigation planning is guided by policies for LENR development and the resulting rate of its deployment. GP—Guidance for mitigation planning.

5.2. Integration of LENR policies among agencies, nations, and the private sector

A focus on integrated LENR policymaking results in opportunities in several other policy areas. Public agency policy integration (PA) may take place at the local, state, and national levels and requires alignment and effective communication of the policymaking entities within the agencies. Formal arrangements, such as inter-agency agreements, may be used, or integration may be achieved by informal measures, such as regular inter-agency meetings. While these measures have been used to some extent by agencies for various issues in the past, they may become increasingly important as LENR deployment progresses.

LENR development—and dealing with its impacts—may be enhanced with stronger integration between the public and private components of society (PP). For example, LENR may benefit from government policies and measures to address "market failures," in a similar vein to current laws and regulations for environmental protection. Existing programs, such as small-business research support and provisions for technology transfer from government labs to privately held companies, could increase in importance if the government becomes more active in LENR research. Public-private partnerships (PPPs) may provide another vehicle for supporting LENR development and realization. An improved stance among patent and trademark entities would also substantially enhance efforts in the private sector to realize the benefits of LENR. Opportunities may be found for integrating these policy changes and updates in the public and private aspects of LENR development.

At the international level, programs may be established for supporting LENR research (IN). As LENR reaches the stage of worldwide deployment, bilateral and multi-lateral agreements may be made or updated to enhance its availability. For example, the United Nations may implement programs for making small LENR units available in a dispersed manner in Third World nations. World Bank loans may be made to nations needing support in acquiring LENR technology for the benefit of human health and the environment. The World Trade Organization may consider LENR and its humanitarian benefits for special rulemaking to enhance availability worldwide. Again, opportunities may be found for integration of policy changes or updates among these international entities.

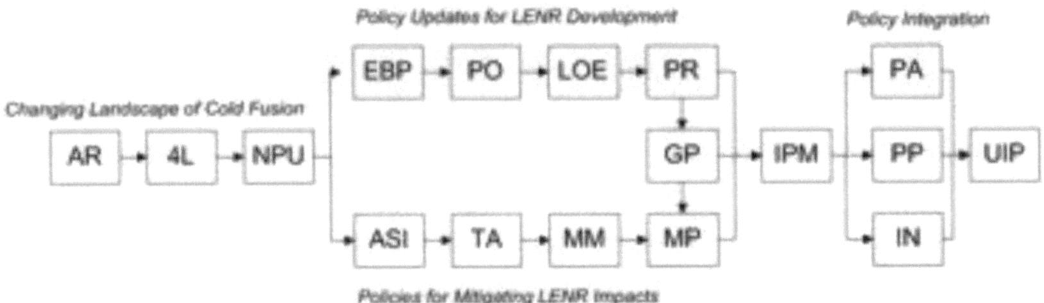

Figure 4. Path to updated and integrated LENR policies. IPM—Integrated policymaking framework; PA—Integration among public agencies; PP—Policy coordination between public and private sectors; IN—Integration among nations at the international level; UIP—Updated and integrated LENR policies.

5.3. Summary path to updated and integrated LENR policies

As policies are updated for LENR support and for mitigating adverse secondary impacts, and as they are integrated at various levels, the public interest will be served for the humanitarian benefits of LENR. **Figure 4** illustrates the full path from the present situation of LENR's changing landscape to the prospective future of fully updated and integrated LENR policies. Integrated policymaking (IPM) for LENR development and mitigating its impacts provides the basis for further updates and integration for public agency (PA), public-private (PP), and international (IN) policymaking. The desired result is fully updated and integrated policies (UIP) for LENR support and impact mitigation as well as among policymaking entities at various levels.

6. Benefits and challenges of LENR policy integration

Achieving integration of LENR policies as updates are accomplished will have substantial public interest benefits. But many challenges must be overcome as well. LENR policy integration will help avoid conflicts and actions that are at cross-purposes among interested parties. Correspondingly, there will be increased efficiency in achieving the policy objectives of the parties as well as increased cost effectiveness where entities have common interests. Policy integration may also achieve improved social equity, with more rational decisions and less influence of purely political considerations.

A principal challenge for policy update and integration is the historical barrier to LENR acceptance after its initial rejection. This barrier seems likely to be surmounted as LENR continues to be investigated and the evidence continues to show that it is real—and that its potential benefits are attainable. Another challenge is the sheer immensity of the expected direct and indirect secondary impacts. Proactive planning to address these impacts will be a major undertaking. The existence of the long-standing and well-established energy policy framework may also present a barrier to effective LENR policies and their integration. This framework includes many conflicting interests and agendas that will have to be considered as policies are updated and integrated.

7. Summary and conclusions

Despite being rejected by mainstream science not long after it was announced, LENR has continued to be pursued in many venues, resulting in improved prospects and the need for policy updates. Updates are needed both for support of LENR development and preparation to mitigate its anticipated adverse secondary impacts. As these updates are accomplished, there are opportunities to integrate the policies to support and realize LENR with mitigation planning for anticipated impacts. There are also policy integration opportunities among public and private entities and at many levels within nations and internationally. The benefits of updating and integrating LENR policies are substantial, but the challenges for doing so are also very large. Opportunities for policy updates and integration may be set forth conceptually, but realization in the "real world" will be much more difficult. Nevertheless, it is clear that the public interest will be served by updating LENR policies and achieving their integration.

Note

The purpose of this Note is to provide attribution for the original source of the paper. It may be deleted if required for this chapter to be included in the book.

Author details

Thomas W. Grimshaw

Address all correspondence to: thomaswgrimshaw@gmail.com

The University of Texas at Austin, Austin, Texas, USA

References

[1] International Society of Condensed Matter Nuclear Science. Date unknown. Online. Available: http://www. iscmns.org/. [Accessed: April 2018]

[2] The 21st International Conference on Cold Fusion (ICCF-21), 3-8 June 2018 Fort Collins, Colorado. Online. Available: https://www.iccf21.com/. [Accessed: April 2018]

[3] Wrbanek S, et al. NASA Glenn Research Center Experience with "LENR Phenomenon", Presentation at Interagency Advanced Power Group (IAPG), Mechanical Working Group (MWG); May 2012

[4] Douglas P, Wells D, et al. Low Energy Nuclear Reaction Aircraft – 2013 ARMD Seedling Fund Phase I Project. Hampton, Virginia. NASA/TM–2014-218283: Langley Research Center; 2013

[5] Defense Intelligence Agency, Worldwide Research on Low-Energy Nuclear Reactions Increasing and Gaining Acceptance. Defense Analysis Report, DIA-08-0911-003. 13 November 2009

[6] LENRIA - The Industrial Association for LENR. Online. Available: https://www.lenria.org/. [Accessed: April 2018]

[7] Anthropocene Institute. LENRaries – A New Era a Renewable Energy. Unpublished Report. 2017. Online. Available: www.anthropoceneinstitute.com. [Accessed: April 2018]

[8] LENR-CANR.org, a Library of Papers about Cold Fusion. Online. Available: http://lenr-canr.org/wordpress/?page_id=1213. [Accessed: April 2018]

[9] Storms E. Science of Low Energy Nuclear Reaction: A Comprehensive Compilation of Evidence and Explanations about Cold Fusion. Singapore: World Scientific Publishing. Tables 2, 6, 8, 11; 2007

[10] Storms E, Grimshaw T. Judging the validity of the Fleischmann-pons effect. The Journal of Condensed Matter Nuclear Science Electronic. 2010;**3**:9-30

[11] Langmiur I. Colloquium on pathological science. Held at the knolls research laboratory, 18 December 1953. Reproduced as "Pathological Science", 1989. Physics Today. **42**(10):36-48

[12] Sagan C. The fine art of baloney detection. Chapter 13. In: The Demon-Haunted World: Science as a Candle in the Dark. New York, NY: Random House; 1995

[13] Shermer M. The Borderlands of Science – Where Sense Meets Nonsense. Oxford, UK: Oxford Univ.; 2001

[14] Hagelstein P, Chaudhary I. Phonon models for anomalies in condensed matter nuclear science. Current Science. 2015;**108**(4):507-513

[15] Storms E. The Evaluation of Low Energy Nuclear Reaction: An Explanation of the Relationship between Observation and Explanation. Concord, NH: Infinite Energy Press; 2014

[16] Hubler G. Sidney Kimmel institute for nuclear renaissance. Current Science. 2015; **108**(4):562-564

[17] Scarbrough T, et al. The center to study the anomalous heat effects. Poster Presented at ICCF-19. Padua, Italy; April 2015

[18] Iwamura Y, et al. The launch of a new plan on condensed matter nuclear science at Tohoku University. Paper Presented at ICCF-19. Padua, Italy. April 2015

[19] Hambling D. Success for Andrea Rossi's E-Cat Cold Fusion System, but Mysteries Remain. San Francisco: Wired (Magazine); 29 October, 2011

[20] Levi G, et al. Indication of Anomalous Heat Energy Production in a Reactor Device Containing Hydrogen Loaded Nickel Powder. Cornell University Library arXiv; 2013

[21] Levi G, et al. Observation of abundant heat production from a reactor device and of isotopic changes of fuel. Unpublished Manuscript; 2014

[22] PRWire, PRNewswire, Industrial Heat Has Acquired Andrea Rossi's E-Cat Technology. Research Triangle Park; 2014

[23] Parkhomov A, Belousova E. Researches of the heat generator similar to high temperature Rosssi reactor. Poster Presented at ICCF-19. Padua, Italy; April 2015

[24] McKubre M. A Russian experiment: High temperature, nickel, natural hydrogen. Infinite Energy. March/April 2015;(120):9-11

[25] Jiang S. New Result of Anomalous Heat Production in Hydrogen-Nickel Metals at High Temperature. Presentation Posted at E-Cat World. www.e-catworld; 2015

[26] JET Energy, Inc. JET Energy Clean Energy Technologies. Online. Available: http://world.std.com/~mica/jettechnology.htm. [Accessed: April 2018]

[27] Hagelstein P. On theory and science generally in connection with the Fleischmann-pons experiment. Infinite Energy. 2013;**108**(March/April):5-12

[28] George R. et al. About Brillouin Corp. Online. Available: http://brillouinenergy.com/about/. [Accessed April 2018]

[29] Brillouin Energy, HHTTM Reactor and Results. Poster at ICCf-19 Conference. Padua, Italy; April 2015

[30] Grimshaw T. Evidence-Based Public Policy toward Cold Fusion: Rational Choices for a Potential Alternative Energy Source. Austin, TX. Lyndon B: Johnson School of Public Affiars. Unpublished Professional Report; 2008

[31] Grimshaw T. Evidence-based public policy for support of cold fusion (LENR) development. Poster Presented at 17th International Conference on Cold Fusion, Daejeon, South Korea; August 2012

[32] Bower J, Christenson C. Disruptive Technologies: Catching the Wave. Watertown: Harvard Business Review; January-February 1995

[33] Christiansen C. The Innovator's Dilemma: When New Technologies Cause Great Firms to Fail. Cambridge, MA: Harvard University Press; 2000

[34] Grimshaw T. Public policy planning for broad deployment of cold fusion (LENR) for energy production. Paper FrM1-1. 17th International Conference on Cold Fusion, Daejeon, South Korea; August 2012

[35] Grimshaw T. Cold Fusion Public Policy: Rational – and Urgent – Need for Change. Presentation at 2014 Cold Fusion/LANR Colloquium at MIT. Cambridge, MA; March 2014

[36] White, Irvin L, et al. Energy from the West: Summary Report. U.S. Environmental Protection Agency. Science and Public Policy Program, University of Oklahoma. Prepared for Office of Research and Development. EPA (600/9-79-027); August 1979

[37] Johns L, et al. A Technology Assessment of Coal Slurry Pipelines. Office of Technology Assessment. NTIS Order #PB-278675; March 1978